Troubleshooting Docker

Strategically design, troubleshoot, and automate Docker
containers from development to deployment

Vaibhav Kohli
Rajdeep Dua
John Wooten

BIRMINGHAM - MUMBAI

Troubleshooting Docker

First published: March 2017

Production reference: 1280317

Published by Packt Publishing Ltd.
Livery Place
35 Livery Street
Birmingham
B3 2PB, UK.

ISBN 978-1-78355-234-4

www.packtpub.com

Credits

Authors

Vaibhav Kohli
Rajdeep Dua
John Wooten

Reviewer

Radhakrishna Nayak

Commissioning Editor

Vedika Naik

Acquisition Editor

Tushar Gupta

Content Development Editors

Johann Barretto
Divij Kotian

Technical Editors

Murtaza Tinwala
Sushant Nadkar

Copy Editors

Shaila Kusanale
Safis Editing

Project Coordinator

Ritika Manoj

Proofreader

Safis Editing

Indexer

Tejal Daruwale Soni

Production Coordinator

Aparna Bhagat

Graphics

Jason Monteiro

About the Authors

Vaibhav Kohli at present is working in VMware's R&D Department and earlier taught computer engineering for a year at the esteemed University of Mumbai. He works for the office of the CTO, VMware IoT (Internet of Things) project. He has published many research papers in leading journals, IEEE transactions, and has filed patents at VMware on container technology. One of his big data projects has won the top prize at the national-level project showcase event. He has conducted workshops, hackathons, training sessions, and trade shows in many countries and is a reputed & established speaker at conferences on IoT and Docker technology. He is an active open source code contributor, repository manager and has also published many online Docker & Kubernetes tutorials. He has helped many customers and organizations to understand cloud-native apps, DevOps model and migrate to micro-service architecture. He has also recently published a book on Docker Networking.

Vaibhav manages and leads various meetup groups across India on the latest cutting-edge Docker and Kubernetes technologies.

> *First and foremost, I would like to thank God for providing me the gift of writing. I dedicate this book to my mom & dad, Kamni Kohli & Ashok K Kohli, who have supported me in everything I have ever done. A special thanks to all my mentors who helped shape me into the person that I am…*

Rajdeep Dua has over 18 years of experience in the Cloud and Big Data space. He has worked extensively on Cloud Infrastructure and Machine Learning related projects as well evangelized these stacks at Salesforce, Google, VMWare. Currently, he leads Developer Relations team at Salesforce India. He also works with the Machine Learning team at Salesforce.

He has been nominated by Docker as a Docker Captain for his contributions to the community. His contributions to the open source community are in the following projects: Docker, Kubernetes, Android, OpenStack, and Cloud Foundry. He also has teaching experience in Cloud and Big Data at IIIT Hyderabad, ISB, IIIT Delhi, and College of Engineering Pune.

Rajdeep did his MBA in IT and Systems from Indian Institute of Management Lucknow, India, and BTech from Thapar University, Patiala, India. You can reach him on Twitter at `@rajdeepdua`.

John Wooten is the founder and CEO of CONSULTED, a global open source cloud consultancy that designs secure, succinct, and sustainable cloud architectures. As a leading cloud solutions architect and open technology strategist, John has deep and applicable real-world experience in designing, testing, deploying, and managing public, private, and hybrid cloud systems for both enterprise and government. His primary technical proficiencies include Linux systems administration, OpenStack clouds, and Docker containers. As an open source activist, John is committedly passionate and works extensively within a broad range of open source projects and communities. His latest Internet project is as the founder and maintainer of ÖppenSourced (`www.öppensourced.com`), a leading repository and defining resource on open source projects and applications. Otherwise, he is a self-proclaimed beach bum in search of the next surfable wave or is hiking and camping in remote locations.

About the Reviewer

Radhakrishna Nayak is a free software evangelist and works as a senior software engineer at Plivo. He has over 5 years of experience. He is a full stack developer and has worked on various open source technologies, such as Python, Django, AngularJS, ReactJS, HTML5, CSS3, MySQL, MongoDB, PostgreSQL, Docker, and many more. He loves to explore and try out new technologies. He is a foodie and loves to explore new restaurants and ice cream parlors.

I would like to thank my parents, Sudheer Nayak and Vranda Nayak, for all their support and freedom to follow my passion. Thanks to all my colleagues at Plivo and Gautham Pai, founder of Jnaapti, for all their support and encouragement. Thanks to Nikhil, Ritika, and the team at Packt Publishing for trusting my abilities and giving the an opportunity to review this book.

www.PacktPub.com

For support files and downloads related to your book, please visit www.PacktPub.com.

Did you know that Packt offers eBook versions of every book published, with PDF and ePub files available? You can upgrade to the eBook version at www.PacktPub.com and as a print book customer, you are entitled to a discount on the eBook copy. Get in touch with us at service@packtpub.com for more details.

At www.PacktPub.com, you can also read a collection of free technical articles, sign up for a range of free newsletters and receive exclusive discounts and offers on Packt books and eBooks.

https://www.packtpub.com/mapt

Get the most in-demand software skills with Mapt. Mapt gives you full access to all Packt books and video courses, as well as industry-leading tools to help you plan your personal development and advance your career.

Why subscribe?

- Fully searchable across every book published by Packt
- Copy and paste, print, and bookmark content
- On demand and accessible via a web browser

Customer Feedback

Thanks for purchasing this Packt book. At Packt, quality is at the heart of our editorial process. To help us improve, please leave us an honest review on this book's Amazon page at https://www.amazon.com/Troubleshooting-Docker-John-Wooten/dp/1783552344.

If you'd like to join our team of regular reviewers, you can e-mail us at customerreviews@packtpub.com. We award our regular reviewers with free eBooks and videos in exchange for their valuable feedback. Help us be relentless in improving our products!

Table of Contents

Preface

Docker is an open source, container-based platform that enables anyone to consistently develop and deploy stable applications anywhere. Docker delivers speed, simplicity, and security in creating scalable and portable environments for ultramodern applications. With the advent and prevalence of Docker in the containerization of modern microservices and N-tier applications, it is both prudent and imperative to effectively troubleshoot automated workflows for production-level deployments.

What this book covers

Chapter 1, *Understanding Container Scenarios and an Overview of Docker*, is about the basic containerization concept with the help of application and OS-based containers. We will throw some light on the Docker technology, its advantages, and the life cycle of Docker containers.

Chapter 2, *Docker Installation*, will go over the steps to install Docker on various Linux distributions – Ubuntu, CoreOS, CentOS, Red Hat Linux, Fedora, and SUSE Linux.

Chapter 3, *Building Base and Layered Images*, teaches that a mission-critical task in production-ready application containerization is image building. We will also discuss building images manually from scratch. Moving ahead, we will explore building layered images with a Dockerfile and enlist the Dockerfile commands in detail.

Chapter 4, *Devising Microservices and N-Tier Applications*, will explore example environments designed seamlessly from development to test, eliminating the need for manual and error-prone resource provisioning and configuration. In doing so, we will touch briefly on how a microservice applications can be tested, automated, deployed, and managed.

Chapter 5, *Moving Around Containerized Application*, will take a look at Docker registry. We will start with the basic concepts of Docker public repository using Docker Hub and the use case of sharing containers with a larger audience. Docker also provides the option to deploy a private Docker registry, which we look into, that can be used to push, pull, and share the Docker containers internally in the organization.

Chapter 6, *Making Containers Work,* will teach you about privileged containers, which have access to all the host devices, and super-privileged containers, which show that the containers can run a background service that can be used to run services in Docker containers to manage the underlying host.

Chapter 7, *Managing the Networking Stack of a Docker Container,* will explain how Docker networking is powered with Docker0 bridge and its troubleshooting issues and configuration. We will also look at troubleshooting the communication issues between Docker networks and the external network. We look into containers communication across multiple hosts using different networking options, such as Weave, OVS, Flannel, and Docker's latest overlay network. We will compare them and look at the troubleshooting issues involved in their configuration.

Chapter 8, *Managing Docker Containers with Kubernetes,* explains how to manage Docker containers with help of Kubernetes. We will cover many deployment scenarios and troubleshooting issues while deploying Kubernetes on a Bare Metal machine, AWS, vSphere, or using minikube. We will also look at deploying Kubernetes pods effectively and debugging Kubernetes issues.

Chapter 9, *Hooking Volume Baggage,* will dive into data volumes and storage driver concepts related to Docker. We will discuss troubleshooting the data volumes with the help of the four approaches and look at their pros and cons. The first case of storing data inside the Docker container is the most basic case, but it doesn't provide the flexibility to manage and handle data in the production environment. The second and third cases are about storing the data using data-only containers or directly on the host. The fourth case is about using a third-party volume plugin, Flocker or Convoy, which stores the data in a separate block and even provides reliability with data, even if the container is transferred from one host to another or if the container dies.

Chapter 10, *Docker Deployment in a Public Cloud - AWS and Azure,* outlines Docker deployment on the Microsoft Azure and AWS public clouds.

What you need for this book

You will need Docker 1.12+ installed on Windows, Mac OS,or Linux machines. Public cloud accounts of AWS, Azure and GCE might be required, which are mentioned in the respective sections of the chapters.

Who this book is for

This book is intended to help seasoned solutions architects, developers, programmers, system engineers, and administrators troubleshoot common areas of Docker containerization. If you are looking to build production-ready Docker containers for automated deployment, you will be able to master and troubleshoot both the basic functions and the advanced features of Docker. Advanced familiarity with the Linux command line syntax, unit testing, the Docker registry, GitHub, and leading container hosting platforms and Cloud Service Providers (CSP) are the prerequisites. In this book you will also learn about ways and means to avoid troubleshooting in the first place.

Conventions

In this book, you will find a number of text styles that distinguish between different kinds of information. Here are some examples of these styles and an explanation of their meaning.

Code words in text, database table names, folder names, filenames, file extensions, pathnames, dummy URLs, user input, and Twitter handles are shown as follows: "Restart the cluster using the `start_k8s.sh` shell script."

A block of code is set as follows:

```
ENTRYPOINT /usr/sbin/sshd -D
VOLUME ["/home"]
EXPOSE 22
EXPOSE 8080
```

Any command-line input or output is written as follows:

```
Docker build -t username/my-imagename -f /path/Dockerfile
```

New terms and **important words** are shown in bold. Words that you see on the screen, for example, in menus or dialog boxes, appear in the text like this: "Specify the **Stack name**, **KeyPair**, and cluster 3."

 Warnings or important notes appear in a box like this.

Tips and tricks appear like this.

Reader feedback

Feedback from our readers is always welcome. Let us know what you think about this book-what you liked or disliked. Reader feedback is important for us as it helps us develop titles that you will really get the most out of. To send us general feedback, simply e-mail feedback@packtpub.com, and mention the book's title in the subject of your message. If there is a topic that you have expertise in and you are interested in either writing or contributing to a book, see our author guide at www.packtpub.com/authors.

Customer support

Now that you are the proud owner of a Packt book, we have a number of things to help you to get the most from your purchase.

Downloading the example code

You can download the example code files for this book from your account at http://www.packtpub.com. If you purchased this book elsewhere, you can visit http://www.packtpub.com/support and register to have the files e-mailed directly to you.

You can download the code files by following these steps:

1. Log in or register to our website using your e-mail address and password.
2. Hover the mouse pointer on the **SUPPORT** tab at the top.
3. Click on **Code Downloads & Errata**.
4. Enter the name of the book in the **Search** box.
5. Select the book for which you're looking to download the code files.
6. Choose from the drop-down menu where you purchased this book from.
7. Click on **Code Download**.

Once the file is downloaded, please make sure that you unzip or extract the folder using the latest version of:

- WinRAR / 7-Zip for Windows
- Zipeg / iZip / UnRarX for Mac
- 7-Zip / PeaZip for Linux

The code bundle for the book is also hosted on GitHub at `https://github.com/PacktPublishing/Troubleshooting-Docker`. We also have other code bundles from our rich catalog of books and videos available at `https://github.com/PacktPublishing/`. Check them out!

Errata

Although we have taken every care to ensure the accuracy of our content, mistakes do happen. If you find a mistake in one of our books-maybe a mistake in the text or the code-we would be grateful if you could report this to us. By doing so, you can save other readers from frustration and help us improve subsequent versions of this book. If you find any errata, please report them by visiting `http://www.packtpub.com/submit-errata`, selecting your book, clicking on the **Errata Submission Form** link, and entering the details of your errata. Once your errata are verified, your submission will be accepted and the errata will be uploaded to our website or added to any list of existing errata under the Errata section of that title.

To view the previously submitted errata, go to `https://www.packtpub.com/books/content/support` and enter the name of the book in the search field. The required information will appear under the **Errata** section.

Piracy

Piracy of copyrighted material on the Internet is an ongoing problem across all media. At Packt, we take the protection of our copyright and licenses very seriously. If you come across any illegal copies of our works in any form on the Internet, please provide us with the location address or website name immediately so that we can pursue a remedy.

Please contact us at `copyright@packtpub.com` with a link to the suspected pirated material.

We appreciate your help in protecting our authors and our ability to bring you valuable content.

Questions

If you have a problem with any aspect of this book, you can contact us at questions@packtpub.com, and we will do our best to address the problem.

1
Understanding Container Scenarios and an Overview of Docker

Docker is one of the most recent successful open source projects which provides the packaging, shipping, and running of any application as lightweight containers. We can actually compare Docker containers to shipping containers that provide a standard, consistent way of shipping any application. Docker is a fairly new project and with the help of this book it will be easy to troubleshoot some of the common problems which Docker users face while installing and using Docker containers.

In this chapter, the emphasis will be on the following topics:

- Decoding containers
- Diving into Docker
- The advantages of Docker containers
- Docker lifecycle
- Docker design patterns
- Unikernels

Decoding containers

Containerization is an alternative to a virtual machine, which involves the encapsulation of applications and providing it with its own operating environment. The basic foundation for containers is **Linux Containers** (**LXC**) which is a user space interface for Linux kernel containment features. With the help of powerful API and simple tools, it lets Linux users create and manage application containers. LXC containers are in-between `chroot` and a fully-fledged virtual machine. Another key difference between containerization and traditional hypervisors is that containers share the Linux kernel used by the operating system running the host machine, thus multiple containers running in the same machine use the same Linux kernel. It gives the advantage of being fast with almost zero performance overhead compared to VMs.

Major use cases of containers are listed in the following sections.

OS containers

OS containers can be easily imagined as a **Virtual Machine** (**VM**) but unlike a VM they share the kernel of the host operating system but provide user space isolation. Similar to a VM, dedicated resources can be assigned to containers and we can install, configure, and run different applications, libraries, and so on, just as you would run on any VM. OS containers are helpful in case of scalability testing where a fleet of containers can be deployed easily with different flavors of distros, which is much less expensive compared to the deployment of VMs. Containers are created from templates or images that determine the structure and contents of the container. It allows you to create a container with the identical environment, same package version, and configuration across all containers mostly used in the case of development environment setups.

There are various container technologies such as LXC, OpenVZ, Docker, and BSD jails which are suitable for OS containers:

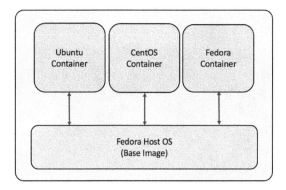

An OS-based container

Application containers

Application containers are designed to run a single service in the package, while OS containers which were explained previously, can support multiple processes. Application containers are attracting a lot of attraction after the launch of Docker and Rocket.

Whenever a container is launched, it runs a single process. This process runs an application process, but in the case of OS containers it runs multiple services on the same OS. Containers usually have a layered approach as in the case of Docker containers, which helps with reduced duplication and increased re-use. Containers can be started with a base image common to all components and then we can go on adding layers in the file system that are specific to the component. A layered file system helps to rollback changes as we can simply switch to old layers if required. The `run` command which is specified in Dockerfile adds a new layer for the container.

The main purpose of application containers is to package different components of the application in separate containers. The different components of the application are packaged separately in containers then they interact with help of APIs and services. The distributed multi-component system deployment is the basic implementation of microservice architecture. In the preceding approach, a developer gets the freedom to package the application as per their requirement and the IT team gets the privilege to deploy the container on multiple platforms in order to scale the system both horizontally as well as vertically:

A hypervisor is a **Virtual Machine Monitor** (**VMM**), used to allow multiple operating systems to run and share the hardware resources from the host. Each virtual machine is termed as a guest machine.

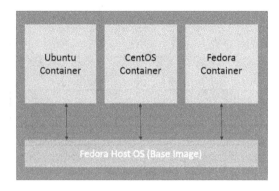

Docker layers

The following simple example explains the difference between application containers and OS containers:

Let's consider the example of web three-tier architecture. We have a database tier such as **MySQL** or **Nginx** for load balancing and the application tier is **Node.js**:

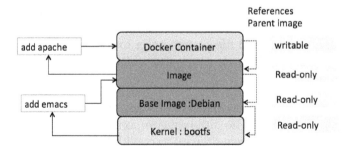

An OS container

In the case of an OS container, we can pick up by default Ubuntu as the base container and install services MySQL, nginx, and Node.js using Dockerfile. This type of packaging is good for a testing or development setup where all the services are packaged together and can be shipped and shared across developers. But deploying this architecture for production cannot be done with OS containers as there is no consideration of data scalability and isolation. Application containers help to meet this use case as we can scale the required component by deploying more application-specific containers and it also helps to meet load balancing and recovery use cases. For the previous three-tier architecture, each of the services will be packaged into separate containers in order to fulfill the architecture deployment use case:

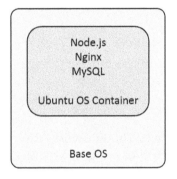

Application containers scaled up

The main differences between OS and application containers are:

OS Container	Application Container
Meant to run multiple services on the same OS container	Meant to run a single service
Natively, no layered filesystem	Layered filesystem
Example: LXC, OpenVZ, BSD Jails	Example: Docker, Rocket

Diving into Docker

Docker is a container implementation that has gathered enormous interest in recent years. It neatly bundles various Linux kernel features and services such as namespaces, cgroups, SELlinux, AppArmor profiles, and so on, with Union file systems such as AUFS and BTRFS to make modular images. These images provide highly configurable virtualized environments for applications and follow the write-once-run-anywhere principle. An application can be as simple as running a process to having highly scalable and distributed processes working together.

Docker is gaining a lot of traction in the industry because of its performance savvy and universally replicable architecture, meanwhile providing the following four cornerstones of modern application development:

- Autonomy
- Decentralization
- Parallelism
- Isolation

Furthermore, wide-scale adaptation of Thoughtworks's microservices architecture or **Lots of Small Applications** (**LOSA**) is further bringing potential to Docker technology. As a result, big companies such as Google, VMware, and Microsoft have already ported Docker to their infrastructure, and the momentum is continued with the launch of myriad of Docker start-ups namely Tutum, Flocker, Giantswarm, and so on.

Since Docker containers replicate their behavior anywhere, be it your development machine, a bare-metal server, virtual machine, or data center, application designers can focus their attention on development, while operational semantics are left to DevOps. This makes team workflow modular, efficient, and productive. Docker is not to be confused with VM, even though they are both virtualization technologies. Where Docker shares an OS, meanwhile providing a sufficient level of isolation and security to applications running in containers, it later completely abstracts out OS and gives strong isolation and security guarantees. But Docker's resource footprint is minuscule in comparison to VM, and hence preferred for economy and performance. However, it still cannot completely replace VM, and the usage of container is complementary to VM technology:

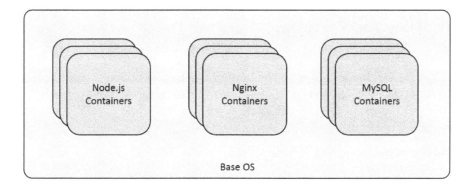

VM and Docker architecture

Advantages of Docker containers

Following are some of the advantages of using Docker containers in microservice architecture:

- **Rapid application deployment**: With minimal runtime, containers can be deployed quickly because of the reduced size as only the application is packaged.
- **Portability**: An application with its operating environment (dependencies) can be bundled together into a single Docker container that is independent from the OS version or deployment model. The Docker containers can be easily transferred to another machine that runs Docker container and executed without any compatibility issues. Windows support is also going to be part of future Docker releases.
- **Easily Shareable**: Pre-built container images can be easily shared with the help of public repositories as well as hosted private repositories for internal use.
- **Lightweight footprint**: Even the Docker images are very small and have a minimal footprint to deploy a new application with the help of containers.
- **Reusability**: Successive versions of Docker containers can be easily built as well as rolled back to previous versions easily whenever required. It makes them noticeably lightweight as components from the pre-existing layers can be reused.

Docker lifecycle

These are some of the basic steps involved in the lifecycle of a Docker container:

1. Build the Docker image with the help of Dockerfile which contains all the commands required to be packaged. It can run in the following way:

   ```
   Docker build
   ```

   Tag name can be added in following way:

   ```
   Docker build -t username/my-imagename .
   ```

   If Dockerfile exists at a different path then the Docker `build` command can be executed by providing `-f` flag:

   ```
   Docker build -t username/my-imagename -f /path/Dockerfile
   ```

2. After the image creation, in order to deploy the container `Docker run` can be used. The running containers can be checked with the help of the `Docker ps` command, which lists the currently active containers. There are two more commands to be discussed:

   - `Docker pause`: This command uses cgroups freezer to suspend all the processes running in a container. Internally it uses the SIGSTOP signal. Using this command process can be easily suspended and resumed whenever required.
   - `Docker start`: This command is used to start one or more stopped containers.

3. After the usage of container is done, it can either be stopped or killed; the `Docker stop:` command will gracefully stop the running container by sending the SIGTERM and then SIGKILL command. In this case, the container can still be listed by using `Docker ps -a` command. `Docker kill` will kill the running container by sending SIGKILL to the main process running inside the container.

4. If there are some changes made to the container while it is running, which are likely to be preserved, a container can be converted back to an image by using the `Docker commit` after the container has been stopped:

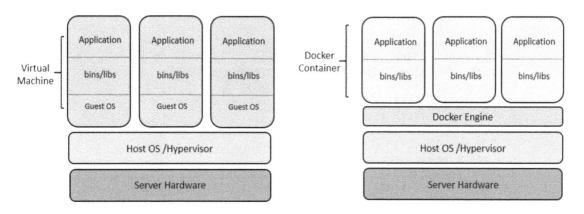

The Docker lifecycle

Docker design patterns

Listed here are eight Docker design patterns with examples. Dockerfile is the base structure from which we define a Docker image, it contains all the commands to assemble an image. Using the `Docker build` command, we can create an automated build that executes all the preceding mentioned command-line instructions to create an image:

```
$ Docker build
Sending build context to Docker daemon 6.51 MB
...
```

Design patterns listed here can help in creating Docker images that persist in volumes and provide various other flexibility so that they can be recreated or replaced easily at any time.

Base image sharing

For creating a web-based application or blog, we can create a base image which can be shared and help to deploy the application with ease. This pattern helps out as it tries to package all the required services on top of one base image, so that this web application blog image can be reused anywhere:

```
FROM debian:wheezy
RUN apt-get update
RUN apt-get -y install ruby ruby-dev build-essential git
# For debugging
RUN apt-get install -y gdb strace
# Set up my user
RUN useradd -u 1000 -ms /bin/bash vkohli
   RUN gem install -n /usr/bin bundler
RUN gem install -n /usr/bin rake
WORKDIR /home/vkohli/
ENV HOME /home/vkohli
VOLUME ["/home"]
USER vkohli
EXPOSE 8080
```

The preceding Dockerfile shows the standard way of creating an application-based image.

 A Docker image is a zipped file which is a snapshot of all the configuration parameters as well as the changes made in the base image (kernel of the OS).

It installs some specific tools (Ruby tools rake and bundler) on top of the Debian base image. It creates a new user, adds it to the container image, and specifies the working directory by mounting "/home" directory from the host, which is explained in detail in the next section.

Shared volume

Sharing the volume at host level allows other containers to pick up the shared content that they require. This helps in faster rebuilding of the Docker image or when adding, modifying, or removing dependencies. For example, if we are creating the homepage deployment of the previously mentioned blog, the only directory required to be shared is the /home/vkohli/src/repos/homepage directory with this web app container through the Dockerfile in the following way:

```
FROM vkohli/devbase
        WORKDIR /home/vkohli/src/repos/homepage
        ENTRYPOINT bin/homepage web
```

For creating the development version of the blog we can share the folder `/home/vkohli/src/repos/blog` where all the related developer files can reside. And for creating the dev-version image we can take the base image from the pre-created `devbase`:

```
FROM vkohli/devbase
WORKDIR /
USER root
# For Graphivz integration
RUN apt-get update
RUN apt-get -y install graphviz xsltproc imagemagick
        USER vkohli
          WORKDIR /home/vkohli/src/repos/blog
          ENTRYPOINT bundle exec rackup -p 8080
```

Development tools container

For development purposes, we have separate dependencies in development and production environments which easily get co-mingled at some point. Containers can be helpful in differentiating the dependencies by packaging them separately. As shown in the following code, we can derive the development tools container image from the base image and install development dependencies on top of it even allowing an `ssh` connection so that we can work upon the code:

```
FROM vkohli/devbase
RUN apt-get update
RUN apt-get -y install openssh-server emacs23-nox htop screen

# For debugging
RUN apt-get -y install sudo wget curl telnet tcpdump
# For 32-bit experiments
RUN apt-get -y install gcc-multilib
# Man pages and "most" viewer:
RUN apt-get install -y man most
RUN mkdir /var/run/sshd
ENTRYPOINT /usr/sbin/sshd -D
VOLUME ["/home"]
EXPOSE 22
EXPOSE 8080
```

As can be seen in the preceding code, basic tools such as `wget`, `curl`, and `tcpdump` are installed which are required during development. Even SSHD service is installed which allows an `ssh` connection into the development container.

Test environment containers

Testing the code in different environments always eases the process and helps find more bugs in isolation. We can create a Ruby environment in a separate container to spawn a new Ruby shell and use it to test the code base:

```
FROM vkohli/devbase
RUN apt-get update
RUN apt-get -y install ruby1.8 git ruby1.8-dev
```

In the Dockerfile listed, we are using the base image as `devbase` and with the help of just one command `docker run` can easily create a new environment by using the image created from this Dockerfile to test the code.

The build container

We have build steps involved in the application that are sometimes expensive. In order to overcome this we can create a separate build container which can use the dependencies needed during the build process. The following Dockerfile can be used to run a separate build process:

```
FROM sampleapp
RUN apt-get update
RUN apt-get install -y build-essential [assorted dev packages for
libraries]
VOLUME ["/build"]
WORKDIR /build
CMD ["bundler", "install","--path","vendor","--standalone"]
```

`/build` directory is the shared directory that can be used to provide the compiled binaries, also we can mount the `/build/source` directory in the container to provide updated dependencies. Thus by using build container we can decouple the build process and the final packaging part in separate containers. It still encapsulates both the process and dependencies by breaking the preceding process into separate containers.

The installation container

The purpose of this container is to package the installation steps in separate containers. Basically, it is in order to provide the deployment of containers in a production environment.

A sample Dockerfile to package the installation script inside a Docker image is shown as follows:

```
ADD installer /installer
CMD /installer.sh
```

The installer.sh can contain the specific installation command to deploy containers in a production environment and also provide the proxy setup with DNS entry in order to have the cohesive environment deployed.

The service-in-a-box container

In order to deploy the complete application in a container, we can bundle multiple services to provide the complete deployment container. In this case we bundle web app, API service, and database together in one container. It helps to ease the pain of interlinking various separate containers:

```
services:
  web:
    git_url: git@github.com:vkohli/sampleapp.git
    git_branch: test
    command: rackup -p 3000
    build_command: rake db:migrate
    deploy_command: rake db:migrate
    log_folder: /usr/src/app/log
    ports: ["3000:80:443", "4000"]
    volumes: ["/tmp:/tmp/mnt_folder"]
    health: default
  api:
    image: quay.io/john/node
    command: node test.js
    ports: ["1337:8080"]
    requires: ["web"]
databases:
  - "mysql"
  - "redis"
```

Infrastructure containers

As we have talked about container usage in a development environment, there is one big category missing–the usage of a container for infrastructure services such as proxy setup which provides a cohesive environment in order to provide the access to an application. In the following mentioned Dockerfile example, we can see that `haproxy` is installed and links to its configuration file are provided:

```
FROM debian:wheezy
ADD wheezy-backports.list /etc/apt/sources.list.d/
RUN apt-get update
RUN apt-get -y install haproxy
ADD haproxy.cfg /etc/haproxy/haproxy.cfg
CMD ["haproxy", "-db", "-f", "/etc/haproxy/haproxy.cfg"]
EXPOSE 80
EXPOSE 443
```

The `haproxy.cfg` file is the configuration file responsible for authenticating a user:

```
backend test
    acl authok http_auth(adminusers)
    http-request auth realm vkohli if !authok
    server s1 192.168.0.44:8084
```

Unikernels

Unikernels compile source code into a custom operating system that includes only the functionality required by the application logic producing a specialized single address space machine image, eliminating unnecessary code. Unikernels are built using the *library operating system*, which has the following benefits compared to a traditional OS:

- **Fast boot time**: Unikernels make provisioning highly dynamic and can boot in less than a second
- **Small footprint**: Unikernel code base is smaller than the traditional OS equivalents and pretty much as easy to manage
- **Improved security**: As unnecessary code is not deployed, the attack surface is drastically reduced
- **Fine-grained optimization**: Unikernels are constructed using compile tool chains and are optimized for device drivers and application logic to be used

Unikernels match very well with the microservices architecture as both source code and generated binaries can be easily version-controlled and are compact enough to be rebuilt. Whereas on the other side, modifying VMs is not permitted and changes can only be made to source code, which is time-consuming and hectic. For example, if the application doesn't require disk access and a display facility. Unikernels can help to remove this unnecessary device driver and display functionality from the kernel. Thus, the production system becomes minimalistic only packaging the application code, runtime environment, and OS facilities which is the basic concept of immutable application deployment where a new image is constructed if any application change is required in production servers:

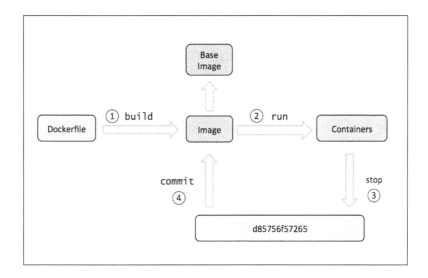

The transition from a traditional container to Unikernel-based containers

Containers and Unikernels are a best fit for each other. Recently, the Unikernel system has become part of Docker and the collaboration of both these technologies will be seen soon in the next Docker release. As explained in the preceding diagram, the first one shows the traditional way of packaging one VM supporting multiple Docker containers. The next step shows a 1:1 map (one container per VM) which allows each application to be self-contained and gives better resource usage, but creating a separate VM for each container adds an overhead. In the last step, we can see the collaboration of Unikernels with the current existing Docker tools and ecosystem, where a container will get the kernel low-library environment specific to its need.

Adoption of Unikernels in the Docker toolchain will accelerate the progress of Unikernels and it will be widely used and understood as a packaging model and runtime framework, making Unikernels another type of container. After the Unikernels abstraction for Docker developers, we will be able to choose either to use a traditional Docker container or the Unikernel container in order to create the production environment.

Summary

In this chapter, we studied the basic containerization concept with the help of application and OS-based containers. The differences between them explained in this chapter will clearly help developers to choose the containerization approach which fits perfectly for their system. We have thrown some light on the Docker technology, its advantages, and the lifecycle of a Docker container. The eight Docker design patterns explained in this chapter clearly show the way to implement Docker containers in a production environment. At the end of the chapter, the Unikernels concept was introduced which is the future of where the containerization domain is moving. In the next chapter, we will be starting with Docker installation troubleshooting issues and its deep dive resolutions.

2
Docker Installation

Docker installation is pretty smooth in most of the operating systems, and there are very few chances of things going wrong. Docker Engine installation is supported mostly on all the Linux, Cloud, Windows, and Mac OS X environments. If the Linux version is not supported, then Docker Engine can be installed using binaries. Docker binary installation is mostly oriented for hackers who want to try out Docker on a variety of OS. It usually involves checking runtime dependencies, kernel dependencies, and using Docker platform-specific binaries in order to move ahead with installation.

Docker Toolbox is an installer, which can be used to quickly install and set up a Docker environment on your Windows or Mac machine. Docker toolbox also installs:

- **Docker client**: It executes commands, such as build and run, and ship containers by communicating with the Docker daemon
- **Docker Machine**: It is a tool used to install Docker Engine on virtual hosts and manages them with the help of Docker Machine commands
- **Docker Compose**: It is a tool used to define and run multicontainer Docker applications
- **Kitematic**: The Docker GUI that runs on Mac OS X and Windows operating system

The installation for Docker with toolbox as well as on various supported OSes is quite straightforward, but nevertheless we have listed potential pitfalls and troubleshooting steps involved.

In this chapter, we explore how to install Docker on various Linux distributions, such as the following:

- Ubuntu
- Red Hat Linux

- CentOS
- CoreOS
- Fedora
- SUSE Linux

All of the above OSes can be deployed on the bare-metal machines, but we have used AWS to deploy in some of the cases, as it's an ideal situation for a production environment. Also, it'll be faster to get the environment up and running in AWS. We have explained the steps for the same in the respective sections in this chapter, which will help you to troubleshoot and speed up the deployment on AWS.

Installing Docker on Ubuntu

Let's get started with installing Docker on Ubuntu 14.04 LTS 64-bit. We can use AWS AMI in order to create our setup. The image can be launched on AMI directly with the help of following link:

```
http://thecloudmarket.com/image/ami-a21529cc-ubuntu-images-hvm-ssd-ubuntu-
trusty-14-04-amd64-server-20160114-5
```

The following diagram illustrates the installation steps required to install Docker on Ubuntu 14.04 LTS:

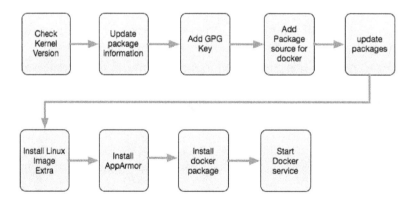

Prerequisites

Docker requires a 64-bit installation, regardless of the Ubuntu version. The kernel must be 3.10 at minimum.

Let's check our kernel version, using the following command:

```
$ uname -r
```

The output is a kernel version of 3.13.x, which is fine:

```
3.13.0-74-generic
```

Updating package information

Perform the following steps to update the APT repository and have necessary certificates installed:

1. Docker's APT repository contains Docker 1.7.x or higher. To set APT to use packages from the new repository:

   ```
   $ sudo apt-get update
   ```

2. Run the following command to ensure that APT works with the HTTPS method and CA certificates are installed:

   ```
   $ sudo apt-get install apt-transport-https ca-certificates
   ```

The `apt-transport-https` package enables us to use `deb https://foo distro main` lines in the `/etc/apt/sources.list` so that package managers, which use the `libapt-pkg` library, can access metadata and packages available in sources accessible over HTTPS.

The `ca-certificates` are container's PEM files of CA certificates, which allow SSL-based applications to check for the authenticity of SSL connections.

Adding a new GPG key

GNU Privacy Guard (known as **GPG** or **GnuPG)** is a free encryption software that's compliant with the OpenPGP (RFC4880) standard:

```
$ sudo apt-key adv --keyserver hkp://p80.pool.sks-keyservers.net:80 --recv-keys 58118E89F3A912897C070ADBF76221572C52609D
```

The output will be similar to the following listing:

```
Executing: gpg --ignore-time-conflict --no-options --no-default-keyring --
homedir /tmp/tmp.SaGDv5OvNN --no-auto-check-trustdb --trust-model always --
keyring /etc/apt/trusted.gpg --primary-keyring /etc/apt/trusted.gpg --
keyserver hkp://p80.pool.sks-keyservers.net:80 --recv-keys
58118E89F3A912897C070ADBF76221572C52609D
gpg: requesting key 2C52609D from hkp server p80.pool.sks-keyservers.net
gpg: key 2C52609D: public key "Docker Release Tool (releasedocker)
<docker@docker.com>" imported
gpg: Total number processed: 1
gpg:                  imported: 1  (RSA: 1)
```

Troubleshooting

If you find the `sks-keyservers` to be unavailable, you can try the following command:

```
$ sudo apt-key adv --keyserver hkp://pgp.mit.edu:80 --recv-keys
58118E89F3A912897C070ADBF76221572C52609D
```

Adding a new package source for Docker

The Docker repository can be added in the following way to the APT repository:

1. Update /etc/apt/sources.list.d with a new source as Docker repository.
2. Open the /etc/apt/sources.list.d/docker.list file and update it with the following entry:

```
deb https://apt.dockerproject.org/repo ubuntu-trusty main
```

Updating Ubuntu packages

The Ubuntu packages after adding Docker repository can be updated, as shown here:

```
$ sudo apt-get update
```

Install linux-image-extra

For Ubuntu Trusty, it's recommended to install the `linux-image-extra` kernel package; the `linux-image-extra` package allows the AUFS storage driver to be used:

```
$ sudo apt-get install linux-image-extra-$(uname -r)
```

The output will be similar to the following listing:

```
Reading package lists... Done
Building dependency tree
Reading state information... Done
The following extra packages will be installed:
  crda iw libnl-3-200 libnl-genl-3-200 wireless-regdb
The following NEW packages will be installed:
  crda iw libnl-3-200 libnl-genl-3-200 linux-image-extra-3.13.0-74-generic
  wireless-regdb
0 upgraded, 6 newly installed, 0 to remove and 70 not upgraded.
Need to get 36.9 MB of archives.
After this operation, 152 MB of additional disk space will be used.
Do you want to continue? [Y/n] Y
Get:1 http://ap-northeast-1.ec2.archive.ubuntu.com/ubuntu/ trusty/main
libnl-3-200 amd64 3.2.21-1 [44 ..
Updating /boot/grub/menu.lst ... done
run-parts: executing /etc/kernel/postinst.d/zz-update-grub 3.13.0-74-
generic /boot/vmlinuz-3.13.0-74-generic
Generating grub configuration file ...
Found linux image: /boot/vmlinuz-3.13.0-74-generic
Found initrd image: /boot/initrd.img-3.13.0-74-generic
done
Processing triggers for libc-bin (2.19-0ubuntu6.6) ...
```

Optional - installing AppArmor

If it is not already installed, install AppArmor using the following command:

```
$ apt-get install apparmor
```

The output will be similar to the following listing:

```
sudo: unable to resolve host ip-172-30-0-227
Reading package lists... Done
Building dependency tree
Reading state information... Done
apparmor is already the newest version.
0 upgraded, 0 newly installed, 0 to remove and 70 not upgraded.
```

Docker installation

Let's get started with installing Docker Engine on Ubuntu using the official APT package:

1. Update APT package index:

   ```
   $ sudo apt-get update
   ```

2. Install Docker Engine:

   ```
   $ sudo apt-get install docker-engine
   ```

3. Start the Docker daemon:

   ```
   $ sudo service docker start
   ```

4. Verify that Docker is installed correctly:

   ```
   $ sudo docker run hello-world
   ```

5. This is how the output would look like:

   ```
   Latest: Pulling from library/hello-world
   03f4658f8b78: Pull complete
   a3ed95caeb02: Pull complete
   Digest: sha256:8be990ef2aeb16dbcb9271ddfe2610fa6658d13f6dfb8b
   c72074cc1ca36966a7
   Status: Downloaded newer image for hello-world:latest
   Hello from Docker.
   This message shows that your installation appears to be working
   correctly.
   ```

Installing Docker on Red Hat Linux

Docker is supported on Red Hat Enterprise Linux 7.x. This section provides an overview of installation of Docker using Docker-managed release packages and installation mechanisms. Using these packages ensures that you will be able get the latest release of Docker.

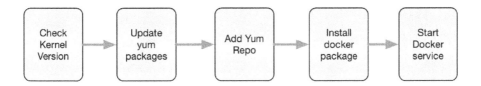

Checking kernel version

The Linux kernel version can be checked with the help of the following command:

```
$ uname -r
```

The output, in our case, is kernel version 3.10.x, which will work fine:

```
3.10.0-327.el7.x86 _64
```

Updating the YUM packages

The YUM repository can be updated, using the following command:

```
$ sudo yum update
```

Output listing is given; ensure that it shows `Complete!` at the end, as follows:

```
Loaded plugins: amazon-id, rhui-lb, search-disabled-repos
rhui-REGION-client-config-server-7          | 2.9 kB
. . . .
Running transaction check
Running transaction test
Transaction test succeeded
Running transaction
  Installing : linux-firmware-20150904-43.git6ebf5d5.el7.noarch      1/138
  Updating   : tzdata-2016c-1.el7.noarch                             2/138
  . . . .
Complete!
```

Adding the YUM repository

Let's add the Docker repository to the YUM repository list:

```
$ sudo tee /etc/yum.repos.d/docker.repo <<-EOF
[dockerrepo]
name=Docker Repository
baseurl=https://yum.dockerproject.org/repo/main/centos/7
enabled=1
gpgcheck=1
gpgkey=https://yum.dockerproject.org/gpg
EOF
```

Installing the Docker package

The Docker Engine can be installed using YUM repository, as follows:

```
$ sudo yum install docker-engine
```

Starting the Docker service

The Docker service can be started with help of the following command:

```
$ sudo service docker start
Redirecting to /bin/systemctl start docker.service
```

Testing the Docker installation

Listing all the processes in the Docker Engine with help of the following command can validate whether the installation of the Docker service is successful or not:

```
$ sudo docker ps -a
```

The following is the output for the preceding command:

```
CONTAINER   ID   IMAGE   COMMAND   CREATED   STATUS   PORTS   NAMES
```

Check the Docker version to make sure that it is the latest:

```
$ docker --version
Docker version 1.11.0, build 4dc5990
```

Checking the installation parameters

Let's run Docker information to see the default installation parameters:

```
$ sudo docker info
```

The output listing is given here; note that the `Storage Driver` is `devicemapper`:

```
Containers: 0
 Running: 0
 Paused: 0
 Stopped: 0
Images: 0
Server Version: 1.11.0
Storage Driver: devicemapper
 Pool Name: docker-202:2-33659684-pool
 Pool Blocksize: 65.54 kB
 Base Device Size: 10.74 GB
 Backing Filesystem: xfs
 Data file: /dev/loop0
 Metadata file: /dev/loop1
 . . .
Cgroup Driver: cgroupfs
Plugins:
 Volume: local
 Network: null host bridge
Kernel Version: 3.10.0-327.el7.x86_64
Operating System: Red Hat Enterprise Linux Server 7.2 (Maipo)
OSType: linux
Architecture: x86_64
CPUs: 1
Total Memory: 991.7 MiB
Name: ip-172-30-0-16.ap-northeast-1.compute.internal
ID: VW2U:FFSB:A2VP:DL5I:QEUF:JY6D:4SSC:LG75:IPKU:HTOK:63HD:7X5H
Docker Root Dir: /var/lib/docker
Debug mode (client): false
Debug mode (server): false
Registry: https://index.docker.io/v1/
```

Troubleshooting tips

Ensure that you are using the latest version of Red Hat Linux to be able to deploy Docker 1.11. Another important thing to remember is that the kernel version has to be 3.10 or higher. Rest of the installation was pretty uneventful.

Deploy CentOS VM on AWS to run Docker containers

We are using AWS as an environment to showcase Docker installation from a convenience perspective. If an OS needs to be tested for support of its Docker version, AWS is the easiest and quickest way to deploy and test it.

If you are not using AWS as an environment, feel free to skip the steps involving spinning a VM on AWS.

In this section, we'll take a look at deploying CentOS VM on AWS to get the environment up and running fast and deploy Docker containers. CentOS is similar to Red Hat's distribution and uses the same packaging tools like YUM. We will use CentOS 7.x, on which Docker is officially supported:

First, let's launch a CentOS-based VM on AWS:

CentOS 7 (x86_64) with Updates HVM

Sold by: Centos.org

This is the Official CentOS 7 x86_64 HVM image that has been built with a minimal profile, suiteable for use in HVM instance types only. The image contains just enough packages to run within AWS, bring up an SSH Server and allow users to login. Please note that this is the default CentOS-7 image that we recommend everyone uses. It contains packages that are updated at points in time to include critical security updates.

Customer Rating	★★★★★ ☑ (42 Customer Reviews)
Latest Version	1602
Operating System	Linux/Unix, CentOS 7
Delivery Method	64-bit Amazon Machine Image (AMI) (Learn more)
Support	See details below
AWS Services Required	Amazon EC2, Amazon EBS

Continue

You will have an opportunity to review your order before launching or being charged.

Pricing Details

For region

Asia Pacific (Tokyo) ⬍

Hourly Fees

Total hourly fees will vary by instance type and EC2 region.

We are launching with a **1-Click Launch** and pre-existing keypair. SSH is enabled by default:

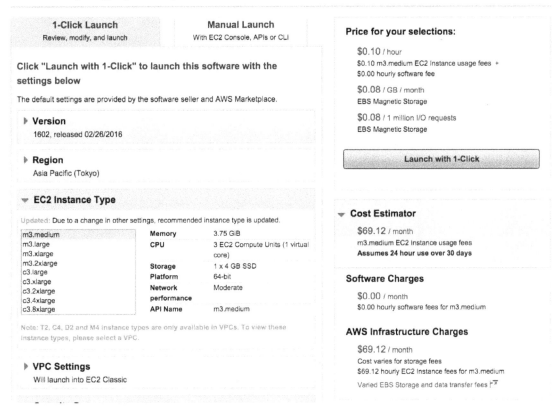

Once the instance is up, get the public IP address from the AWS EC2 console.

SSH into the instance and follow the following steps for installation:

```
$ ssh -i "ubuntu-1404-1.pem" centos@54.238.154.134
```

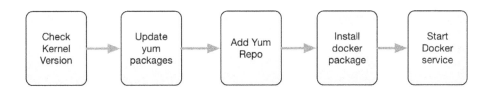

Checking kernel version

The kernel version of the Linux OS can be checked with the following command:

```
$ uname -r
```

The output, in our case, is kernel version 3.10.x, which will work fine:

```
3.10.0-327.10.1.el7.x86_64
```

Note how similar it is to the Red Hat kernel version 3.10.0-327.el7.x86_64.

Updating the YUM packages

The YUM packages and repository can be updated, as shown here:

```
$ sudo yum update
Output listing is given, make sure it shows complete at the end

Loaded plugins: fastestmirror
base                                                   | 3.6 kB     00:00
extras                                                 | 3.4 kB     00:00
updates                                                | 3.4 kB     00:00
(1/4): base/7/x86_64/group_gz                          | 155 kB   00:00
(2/4): extras/7/x86_64/primary_db                      | 117 kB   00:00
(3/4): updates/7/x86_64/primary_db                     | 4.1 MB   00:00
(4/4): base/7/x86_64/primary_db                        | 5.3 MB   00:00
Determining fastest mirrors
 * base: ftp.riken.jp
 * extras: ftp.riken.jp
 * updates: ftp.riken.jp
Resolving Dependencies
--> Running transaction check
---> Package bind-libs-lite.x86_64 32:9.9.4-29.el7_2.2 will be updated
---> Package bind-libs-lite.x86_64 32:9.9.4-29.el7_2.3 will be an update
---> Package bind-license.noarch 32:9.9.4-29.el7_2.2 will be updated
---> Package bind-license.noarch 32:9.9.4-29.el7_2.3 will be an update
. . . .
  teamd.x86_64 0:1.17-6.el7_2
  tuned.noarch 0:2.5.1-4.el7_2.3
  tzdata.noarch 0:2016c-1.el7
  util-linux.x86_64 0:2.23.2-26.el7_2.2
Complete!
```

Adding the YUM repository

Let's add the Docker repository to the YUM repository:

```
$ sudo tee /etc/yum.repos.d/docker.repo <<-EOF
[dockerrepo]
name=Docker Repository
baseurl=https://yum.dockerproject.org/repo/main/centos/7
enabled=1
gpgcheck=1
gpgkey=https://yum.dockerproject.org/gpg
EOF
```

Installing the Docker package

The following command can be used to install Docker Engine using the YUM repository:

```
$ sudo yum install docker-engine
```

Starting the Docker service

The Docker service can be started in the following way:

```
$ sudo service docker start
Redirecting to /bin/systemctl start docker.service
```

Testing the Docker installation

```
$ sudo docker ps -a
```

Output:

```
CONTAINER ID IMAGE COMMAND CREATED STATUS PORTS NAMES
```

Check the Docker version to make sure it is the latest:

```
$ docker --version
Docker version 1.11.0, build 4dc5990
```

Checking the installation parameters

Let's run Docker information to see the default installation parameters:

```
$ sudo docker info
```

The output is listed here; note that the Storage Driver is devicemapper:

```
Server Version: 1.11.0
Storage Driver: devicemapper
 ...
Kernel Version: 3.10.0-327.10.1.el7.x86_64
Operating System: CentOS Linux 7 (Core)
OSType: linux
Architecture: x86_64
CPUs: 1
Total Memory: 991.7 MiB
Name: ip-172-30-0-236
ID: EG2K:G4ZR:YHJ4:APYL:WV3S:EODM:MHKT:UVPE:A2BE:NONM:A7E2:LNED
Docker Root Dir: /var/lib/docker
Registry: https://index.docker.io/v1/
```

Installing Docker on CoreOS

CoreOS is a lightweight OS built for the cloud. It comes prepackaged with Docker, which is a few releases behind the latest version. Since it comes prebuilt with Docker there is little troubleshooting required. We just need to make sure that the right version of CoreOS is picked.

CoreOS runs on a variety of platforms, including Vagrant, Amazon EC2, QEMU/KVM, VMware and OpenStack, and custom hardware. CoreOS uses fleet to manage clusters of containers along with etcd (key value data store).

Installation channels of CoreOS

In our case, we will use stable **Release Channels**:

Release Channels

Stable	899.15.0	Beta	991.2.0	Alpha	1010.1.0
The Stable channel should be used by production clusters. Versions of CoreOS are battle-tested within the Beta and Alpha channels before being promoted.		The Beta channel consists of promoted Alpha releases. Mix a few beta machines into your production clusters to catch any bugs specific to your hardware or configuration.		The Alpha channel closely tracks current development work and is released frequently. The newest versions of docker, etcd and fleet will be available for testing.	
kernel:	4.3.6	kernel:	4.4.6	kernel:	4.5.0
rkt:	1.0.0	rkt:	1.1.0	rkt:	1.2.1
docker:	1.9.1	docker:	1.9.1	docker:	1.10.3
Release Notes • Browse Images		Release Notes • Browse Images		Release Notes • Browse Images	

First, we will install CoreOS on AWS using the CloudFormation templates. You can find this template at the following link:

```
https://s3.amazonaws.com/coreos.com/dist/aws/coreos-stable-pv.template
```

This template provides the following parameters:

- Instance type
- Cluster size
- Discovery URL
- Advertised IP address
- Allow SSH From
- Keypair

These mentioned parameters can be set in the default template, as follows:

```
{
  "Parameters": {
    "InstanceType": {
      "Description": "EC2 PV instance type (m3.medium, etc).",
      "Type": "String",
      "Default": "m3.medium",
      "ConstraintDescription": "Must be a valid EC2 PV instance type."
    },
    "ClusterSize": {
      "Default": "3",
      "MinValue": "3",
```

```
        "MaxValue": "12",
        "Description": "Number of nodes in cluster (3-12).",
        "Type": "Number"
    },
    "DiscoveryURL": {
        "Description": "An unique etcd cluster discovery URL. Grab a new
token from https://discovery.etcd.io/new?size=<your cluster size>",
        "Type": "String"
    },
    "AdvertisedIPAddress": {
        "Description": "Use 'private' if your etcd cluster is within one
region or 'public' if it spans regions or cloud providers.",
        "Default": "private",
        "AllowedValues": [
           "private",
           "public"
        ],
        "Type": "String"
    },
    "AllowSSHFrom": {
        "Description": "The net block (CIDR) that SSH is available to.",
        "Default": "0.0.0.0/0",
        "Type": "String"
    },
    "KeyPair": {
        "Description": "The name of an EC2 Key Pair to allow SSH access to
the instance.",
        "Type": "String"
    }
  }
}
```

The following steps will provide the complete walk-through for CoreOS installation on AWS with help of screenshots:

1. Select the S3 template to launch:

Select Template

Select the template that describes the stack that you want to create. A stack is a group of related resources that you manage as a single unit.

Design a template	Use AWS CloudFormation Designer to create or modify an existing template. Learn more.
	Design template
Choose a template	A template is a JSON-formatted text file that describes your stack's resources and their properties. Learn more.
	○ Select a sample template
	○ Upload a template to Amazon S3
	Choose File No file chosen
	● Specify an Amazon S3 template URL
	https://s3.amazonaws.com/coreos.com/dist/aws/coreos-stable-pv.template View/Edit template in Designer

2. Specify the **Stack name**, **KeyPair**, and cluster 3:

Specify Details

Specify a stack name and parameter values. You can use or change the default parameter values, which are defined in the AWS CloudFormation template. Learn more.

Stack name	CoreOS-stable

Parameters

AdvertisedIPAddress	private	Use 'private' if your etcd cluster is within one region or 'public' if it spans regions or cloud providers.
AllowSSHFrom	0.0.0.0/0	The net block (CIDR) that SSH is available to.
ClusterSize	3	Number of nodes in cluster (3-12).
DiscoveryURL		An unique etcd cluster discovery URL. Grab a new token from https://discovery.etcd.io/new?size=<your cluster size>
InstanceType	m3.medium	EC2 PV instance type (m3.medium, etc).
KeyPair	ubuntu-1404-1	The name of an EC2 Key Pair to allow SSH access to the instance

Troubleshooting

Here are some troubleshooting tips and guidelines, which should be followed during the preceding installation:

- **Stack name** should not be duplicate
- **ClusterSize** cannot be lower than 3 and should be a maximum of 12
- **InstanceType** recommended is `m3.medium`
- **KeyPair** should exist; if it doesn't, the cluster will not launch

SSH into the instance and check the Docker version:

```
core@ip-10-184-155-153 ~ $ docker --version
Docker version 1.9.1, build 9894698
```

Installing Docker on Fedora

Docker is supported on Fedora version 22 and 23. The following are the steps to be performed in order to install Docker on Fedora 23. It can be deployed on bare-metal or as a VM.

Checking Linux kernel Version

Docker requires 64-bit installation, regardless of the Fedora version. Also, the kernel version should be at least 3.10. Check the kernel version before going ahead with installation using the following command:

```
$ uname -r
4.4.7-300.fc23.x86_64
Switch to root user
[os@osboxes ~]# su -
Password:
[root@vkohli ~]#
```

Installing with DNF

Update the existing DNF package by using the following command:

```
$ sudo dnf update
```

Adding to the YUM repository

Let's add the Docker repository to the YUM repository:

```
$ sudo tee /etc/yum.repos.d/docker.repo <<-'EOF'
> [dockerrepo]
> name=Docker Repository
> baseurl=https://yum.dockerproject.org/repo/main/fedora/$releasever/
> enabled=1
> gpgcheck=1
> gpgkey=https://yum.dockerproject.org/gpg
> EOF
[dockerrepo]
name=Docker Repository
baseurl=https://yum.dockerproject.org/repo/main/fedora/$releasever/
enabled=1
gpgcheck=1
gpgkey=https://yum.dockerproject.org/gpg
```

Installing the Docker package

The Docker Engine can be installed using the DNF package:

```
$ sudo dnf install docker-engine
```

The output will be similar to the following listing (this listing is truncated):

```
Docker Repository                          32 kB/s | 7.8 kB     00:00
Last metadata expiration check: 0:00:01 ago on Thu Apr 21 15:45:25 2016.
Dependencies resolved.
Install  7 Packages
...
Running transaction test
Transaction test succeeded.
Running transaction
  Installing: python-IPy-0.81-13.fc23.noarch
....
Installed:
...
Complete!
```

Start the Docker service using `systemctl`:

```
$ sudo systemctl start docker
```

Verify with a Docker hello-world example in order to check whether Docker is installed successfully:

```
[root@osboxes ~]# docker run hello-world
```

The output will be similar to the following listing:

```
Unable to find image 'hello-world:last' locally
latest: Pulling from library/hello-world
03f4658f8b78: Pull complete
a3ed95caeb02: Pull complete
Digest:
sha256:8be990ef2aeb16dbcb9271ddfe2610fa6658d13f6dfb8bc72074cc1ca36966a7
Status: Downloaded newer image for hello-world:latest

Hello from Docker.
This message shows that your installation appears to be working correctly.
```

To generate this message, Docker took the following steps:

1. The Docker client contacted the Docker daemon.
2. The Docker daemon pulled the `hello-world` image from the Docker Hub.
3. The Docker daemon created a new container from that image, which runs the executable that produces the output you are currently reading.
4. The Docker daemon streamed that output to the Docker client, which sent it to your terminal.

To try something more ambitious, you can run an Ubuntu container with the following command:

```
$ docker run -it ubuntu bash
```

Share images, automate workflows, and more with a free Docker Hub account, `https://hub.docker.com`.

For more examples and ideas, visit `https://docs.docker.com/userguide/md64-server-20160114.5 (ami-a21529cc)`.

Installing Docker with script

Update your DNF package, as follows:

```
$ sudo dnf update
```

Running the Docker installation script

The Docker installation can also be done in a quick and easy way by executing the shell script and getting it from the official Docker website:

```
$ curl -fsSL https://get.docker.com/ | sh
+ sh -c 'sleep 3; dnf -y -q install docker-engine'
```

Start the Docker daemon:

```
$ sudo systemctl start docker
```

Docker run hello-world:

```
$ sudo docker run hello-world
```

To create a Docker group and add a user, follow the steps mentioned, as follows:

```
$ sudo groupadd docker
$ sudo usermod -aG docker your_username
```

Log out and log in with the user to make sure that your user is created successfully:

```
$ docker run hello-world
```

In order to uninstall Docker, follow these steps:

```
# sudo dnf -y remove docker-engine.x86_64
```

The truncated output of the preceding command is listed as follows:

```
Dependencies resolved.
Transaction Summary
================================================================
Remove  7 Packages
Installed size: 57 M
Running transaction check
Transaction check succeeded.
Running transaction test
Transaction test succeeded.
Running transaction
...
Complete!
[root@osboxes ~]# rm -rf /var/lib/docker
```

Installing Docker on SUSE Linux

In this section, we will install Docker on SUSE Linux Enterprise Server 12 SP1 x86_64 (64-bit). We will also look at some of the problems we came across during the installation process.

Launching the SUSE Linux VM on AWS

Choose the appropriate AMI and launch the instance from the EC2 console:

SUSE Linux Enterprise Server 12 SP1 (HVM), SSD Volume Type - ami-f8220896

SUSE Linux Enterprise Server 12 Service Pack 1 (HVM), EBS General Purpose (SSD) Volume Type. Public Cloud, Advanced Systems Management, Web and Scripting, and Legacy modules enabled.

Root device type: ebs   Virtualization type: hvm

The following parameters are shown in the next step; review and then launch them:

We chose an existing keypair to SSH into the instance:

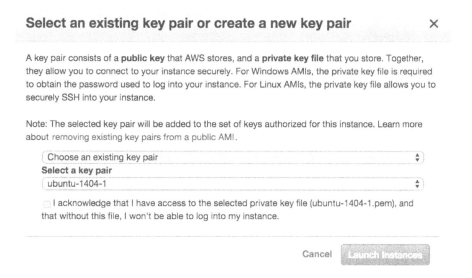

Once the VM is up, log in to the VM from a terminal:

```
$ ssh -i "ubuntu-1404-1.pem" ec2-user@54.199.222.91
```

The truncated output is listed here:

```
The authenticity of host '54.199.222.91 (54.199.222.91)' can't be
established.
...
Management and Config: https://www.suse.com/suse-in-the-cloud-basics
Documentation: http://www.suse.com/documentation/sles12/
Forum: https://forums.suse.com/forumdisplay.php?93-SUSE-Public-Cloud
Have a lot of fun...
ec2-user@ip-172-30-0-104:~>
```

Since we have launched the VM, let's focus on getting docker installed. The following diagram outlines the steps for installing docker on SUSE Linux:

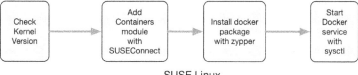

Checking Linux kernel version

Kernel version should be at least 3.10. Check the kernel version before going ahead with its installation, using the following command:

```
$ uname -r
```

Adding Containers-Module

The following Containers-Module needs to be updated in the local packages before docker can be installed. You can find more details about the Containers-Module at the following link:

```
https://www.suse.com/support/update/announcement/2015/suse-ru-20151158-1.html
```

Execute the following command:

```
ec2-user@ip-172-30-0-104:~> sudo SUSEConnect -p sle-module-
containers/12/x86_64 -r ''
```

The output will be similar to this:

```
Registered sle-module-containers 12 x86_64
To server: https://smt-ec2.susecloud.net
ec2-user@ip-172-30-0-104:~>
```

Installing Docker

Execute the following command:

```
ec2-user@ip-172-30-0-104:~> sudo zypper in Docker
```

The truncated output is listed here:

```
...
 (2/2) Installing: docker-1.10.3-66.1
.................................................[done]
Additional rpm output:
creating group docker...
Updating /etc/sysconfig/docker...
```

Starting Docker service

The Docker service can be started, as shown here:

```
ec2-user@ip-172-30-0-104:~> sudo systemctl start docker
```

Checking the Docker installation

Execute Docker run, as follows, to test the installation:

```
ec2-user@ip-172-30-0-104:~> sudo docker run hello-world
```

The output will be similar to this:

```
Unable to find image 'hello-world:latest' locally
latest: Pulling from library/hello-world
4276590986f6: Pull complete
a3ed95caeb02: Pull complete
Digest:
sha256:4f32210e234b4ad5cac92efacc0a3d602b02476c754f13d517e1ada048e5a8ba
Status: Downloaded newer image for hello-world:latest
Hello from Docker.
This message shows that your installation appears to be working correctly.
....
For more examples and ideas, visit:
 https://docs.docker.com/engine/userguide/
ec2-user@ip-172-30-0-104:~>
```

Troubleshooting

Please note, the Docker installation on SUSE Linux 11 is not a smooth experience, as SUSE Connect is not available.

Summary

In this chapter, we went over steps on how to install Docker on various Linux distributions–Ubuntu, CoreOS, CentOS, Red Hat Linux , Fedora, and SUSE Linux. We noticed similarities in the steps and common prerequisites across Linux, while the actual remote repository from where the Docker module needs to be downloaded and the package management for the Docker modules, various for each Linux operating system. In the next chapter, we'll explore the mission-critical task of image building, understanding base and layered images, and exploring the troubleshooting aspect of it.

3
Building Base and Layered Images

In this chapter, we will learn about building base and layered images for production-ready containers. As we saw, Docker containers provide us with ideal environments in which we can build, test, automate, and deploy. The reproductive nature of these exact environments affords a higher degree of efficacy and confidence that currently available script-based deployment systems cannot readily duplicate. The images a developer locally builds, tests, and debugs can then be pushed directly into staging and production environments as the test environment is nearly a mirror image under which the application code runs.

Images are the literal foundational component of containers, defining what flavor of Linux to deploy and what default tools to include and make available to the code running inside the container. Image building is, therefore, one of the most critical tasks in the application containerization life cycle; correctly building your images is critical for effective, repeatable, and secure functionality of containerized applications.

A container image consists of a set of runtime variables for your application container. Ideally, container images should be as minimal as possible, providing the required functionalities only, as this helps in efficient handling of the container image, significantly reducing the time to upload and download the image from the registry and having a minimal footprint on the host.

Our focus, intent, and direction is in building, debugging, and automating images for your Docker containers.

We will cover the following topics in this chapter:

- Building container images
- Building base images from scratch
- Official base images from Docker registry
- Building layered images from Dockerfiles
- Debugging images through testing
- Automated image building with testing

Building container images

As this book attempts to *troubleshoot Docker*, wouldn't it prove beneficial to reduce our chances for errors that we would have to troubleshoot in the first place? Fortunately for us, the Docker community (and the open source community at large) provides a healthy registry of base (or *root*) images that dramatically reduce errors and provide more repeatable processes. Searching the **Docker Registry**, we can find official and automated build statuses for a broad and growing array of container images. The Docker official repositories (`https://docs.docker.com/docker-hub/official_repos/`) are carefully organized collections of images supported by Docker Inc.–automated repositories that allow you to validate source and content of a particular image also exist.

A major thrust and theme of this chapter will be in basic Docker fundamentals; while they may seem trivial to the experienced container user, following some best practices and levels of standardization will serve us well in avoiding trouble spots in addition to enhancing our abilities to troubleshoot.

Official images from the Docker Registry

Standardization is a major component for repeatable processes. As such, wherever and whenever possible, one should opt for a standard base image as provided in the **Docker Hub** for the variant Linux distributions (for example, CentOS, Debian, Fedora, RHEL, Ubuntu, and others) or for specific use cases (for example, WordPress applications). Such base images are derived from their respective Linux platform images, and are built specifically for use in containers. Further, standardized base images are well maintained and updated frequently to address security advisories and critical bug fixes.

These base images are built, validated, and supported by Docker Inc. and are easily recognized by their single word names (for example, `centos`). Additionally, user members of the Docker community also provide and maintain prebuilt images to address certain use cases. Such user images are denoted with the prefix of the Docker Hub username that created them, suffixed with the image name (for example, `tutum/centos`).

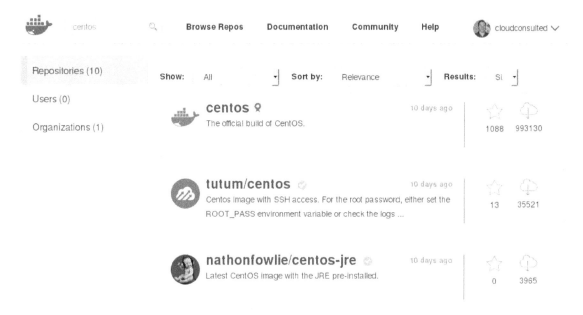

To our great advantage, these standard base images remain ready and are publicly available on the Docker Registry; images can be searched for and retrieved simply using the `docker search` and `docker pull` Terminal commands. These will download any image(s) that are not already located on the Docker host. The Docker Registry has become increasingly powerful in providing official base images for which one can use directly, or at least as a readily available starting point toward addressing the needs of your container building.

While this book assumes your familiarity with Docker Hub/Registry and GitHub/Bitbucket, we will dedicate initial coverage of these as your first line of reference for efficient image building for containers. You can visit the official registry of Docker images at `https://registry.hub.docker.com/`.

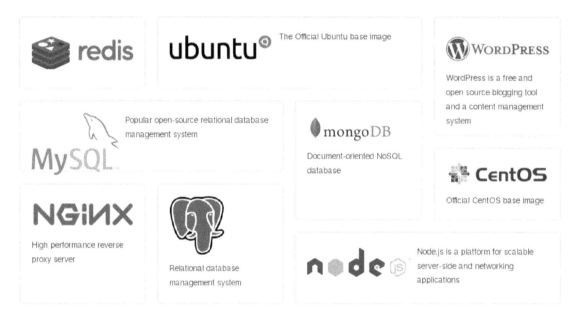

The Docker Registry can be searched from your Docker Hub account or directly from the Terminal, as follows:

```
$ sudo docker search centos
```

Flags can be applied to your search criteria to filter images for star ratings, automated builds, and many more. To use the official `centos` image from the registry, from a Terminal:

- `$ sudo docker pull centos`: This will download the `centos` image to your host machine.
- `$ sudo docker run centos`: This will first look for this image localized on your host and, if not found, it will download the image to host. The run parameters for the image will have been defined in its Dockerfile.

User repositories

Further, as we have seen, we are not limited merely to the repositories of official Docker images. Indeed, a wealth of community users (both as individuals and from corporate enterprises) have prepared images constructed to meet certain needs. As an example, an `ubuntu` image is created to run the `joomla` content management system within a container running on Apache, MySql, and PHP.

Here, we have a user repository with just such an image (`namespace/repository name`):

 Try it out: Practice an image `pull` and `run` from the Docker Registry from the Terminal.
`$ sudo docker pull cloudconsulted/joomla`
pulls our base image for a container and `$ sudo docker run -d -p 80:80 cloudconsulted/joomla`
runs our container image and maps port `80` of the host to port `80` of the container.
Point your browser to `http://localhost` and you will have the build page for a new Joomla website!

Building our own base images

There may be occasion, however, when we need to create custom images to suit our own development and deployment environment. If your use case dictates using a nonstandardized base image, you will need to roll your own image. As with any approach, appropriate planning beforehand is necessary. Before building an image, you should spend adequate time to fully understand the use case your container is meant to address. There isn't much need for a container that cannot run the intended application. Other considerations may include whether the library or binary you are including in the image is reusable, and many more. Once you feel you are done, review your needs and requirements once more and filter out parts that are unnecessary; we do not want to bloat our containers for no good reason.

Using the Docker Registry, you may find automated builds. These builds are pulled from repositories at GitHub/Bitbucket and can, therefore, be forked and modified to your own specifications. Your newly forked repository can then in turn be synced to the Docker Registry with your new image, which can then be pulled and run as needed for your containers.

Try it out: Pull the ubuntu minimal image from the following repository and drop it to your Dockerfile directory to create your own image:
```
$ sudo docker pull cloudconsulted/ubuntu-dockerbase
$ mkdir dockerbuilder
$ cd dockerbuilder
```
Open an editor (vi/vim or nano) and create a new Dockerfile:
```
$ sudo nano Dockerfile
```

We will delve into creating good Dockerfiles later as we talk about layered and automated image building. For now, we just want to create our own new base image, only symbolically going through the procedure and location for creating a Dockerfile. For the sake of simplicity, here we are just calling the base image from which we want to build our new image:

```
FROM cloudconsulted/ubuntu-dockerbase:latest
```

Save and close this Dockerfile. We now build our new image locally:

```
$ sudo docker build -t mynew-ubuntu
```

Let's check to ensure our new image is listed:

```
$ sudo docker images
```

Note our **IMAGE ID** for **mynew-ubuntu**, as we will need it shortly:

Create a new public/private repository under your Docker Hub username. I'm adding the new repository here under <namespace><reponame> as cloudconsulted/mynew-ubuntu:

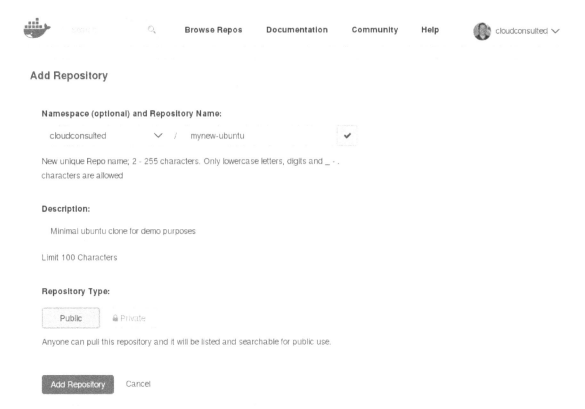

[55]

Next, return to the Terminal so that we can tag our new image to push to the new Docker Hub repository under our `<namespace>`:

```
$ sudo docker tag 1d4bf9f2c9c0 cloudconsulted/mynew-ubuntu:latest
```

Ensure that our new image is correctly tagged for `<namespace><repository>` in our images list:

```
$ sudo docker images
```

Also, we will find our newly created image labeled for pushing it to our Docker Hub repository.

Now, let's push the image up to our Docker Hub repository:

```
$ sudo docker push cloudconsulted/mynew-ubuntu
```

Then, check the Hub for our new image:

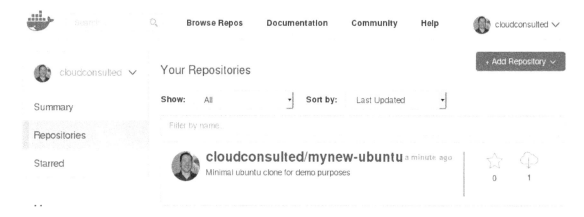

There are essentially two approaches to building your own Docker images:

- Manually constructing layers interactively via bash shell to install necessary applications
- Automating through a Dockerfile that builds the images with all necessary applications

Building images using the scratch repository

Going about building your own container images for Docker is highly dependent on which Linux distribution you intend to package. With such variance, and with the prevalence and growing registry of images already available to us via the Docker Registry, we won't spend much time on such a manual approach.

Here again, we can look in the Docker Registry to provide us with a minimal image to use. A `scratch` repository has been created from an empty TAR file that can be utilized simply via `docker pull`. As before, make your Dockerfile according to your parameters, and you have your new image, from scratch.

This process can be even further simplified by making use of available tools, such as **supermin** (Fedora systems) or **debootstrap** (Debian systems). Using such tools, the build process for an Ubuntu base image, for example, can be as simple as follows:

```
$ sudo debootstrap raring raring > /dev/null
$ sudo tar -c raring -c . | docker import - raring a29c15f1bf7a
$ sudo docker run raring cat /etc/lsb-release
DISTRIB_ID=Ubuntu
DISTRIB_RELEASE=14.04
DISTRIB_CODENAME=raring
DISTRIB_DESCRIPTION="Ubuntu 14.04"
```

Building layered images

A core concept and feature of Docker is layered images. One of the most important features of Docker is **image layering** and the management of image content. A layered approach for container images is very efficient, as you can reference the contents in the image, identifying the layer in a layered image. This is very powerful when building multiple images, using the Docker Registry to push and pull images.

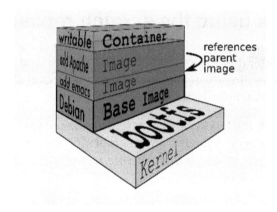

[Image Copyright © Docker, Inc.]

Building layered images using Dockerfiles

Layered images are primarily built using the **Dockerfile**. In essence, a Dockerfile is a script that automatically builds our containers from a source (*base* or *root*) image in the order you need them executed by the Docker daemon, step by step, layer upon layer. These are successive commands (instructions) and arguments enlisted within the file that execute a proscribed set of actions on a base image, with each command constituting a new layer, in order to build a new one. This not only facilitates the organization of our image building but greatly enhances deployments from beginning to end through its simplification. The scripts within a Dockerfile can be presented to the Docker daemon in a range of ways to build new images for our containers.

Dockerfile construction

The first command of a Dockerfile is typically the FROM command. FROM specifies the base image to be pulled. This base image can be located in the public Docker registry (https://www.docker.com/) within a private registry or even a localized Docker image from the host.

Additional layers in a Docker image are populated as per the directives defined in the Dockerfile. Dockerfiles have very handy directives. Every new directive defined in the Dockerfile constitutes a **layer** in a layered image. With a RUN directive, we can specify a command to be run, with the result of the command as an additional layer in the image.

It is highly advised to logically group the operations performed in an image and keep the number of layers to a minimum. For example, while trying to install the dependencies for your application, one can install all the dependencies in one RUN directive rather than using N number of directives per dependency.

We will inspect more closely, the aspects of Dockerfiles for automation in a later section, *Automated image building*. For now, we need to make certain that we grasp the concept and construction of the Dockerfile itself. Let's look specifically at a simple list of commands that can be employed. As we have seen before, our Dockerfile should be created in a working directory containing our existing code (and/or other dependencies, scripts, and others).

CAUTION: Avoid use of the root [/] directory as root of your source repository. The docker build command makes use of the directory containing your Dockerfile as the build context (including all of its subdirectories). The build context will be sent to the Docker daemon before building the image, which means if you use / as the source repository, the entire contents of your hard drive will get sent to the daemon (and thus to the machine running the daemon). In most cases, it is best to put each Dockerfile in an empty directory. Then, only add the files needed for building the Dockerfile to the directory. To increase the build's performance, a .dockerignore file can be added to the context directory to properly exclude files and directories.

Dockerfile commands and syntax

While simplistic, the order and syntax of our Dockerfile commands are extremely important. Proper attention to details and best practice here will not only help ensure successful automated deployments, but also serve to help in any troubleshooting efforts.

Let's delineate some basic commands and illustrate them directly with a working Dockerfile; our joomla image from before is a good example of a basic layered image build from a Dockerfile.

Our sample joomla base image is located in the public Docker index via `cloudconsulted/joomla`.

FROM

A proper Dockerfile begins with defining an image FROM, from which the build process starts. This instruction specifies the base image to be used. It should be the first instruction in Dockerfile, and it is a must for building an image via Dockerfile. You can specify the local image, an image present at the Docker public registry, or image at a private registry.

Common Constructs

```
FROM <image>
FROM <image>:<tag>
FROM <image>@<digest>
```

`<tag>` and `<digest>` are optional; if you do not specify them, it defaults to `latest`.

Example Dockerfile from our Joomla Image

Here, we define the base image to be used for the container:

```
# Image for container base
FROM ubuntu
```

MAINTAINER

This line designates the *Author* of the built image. This is an optional instruction in Dockerfile; however, one should specify this instruction with the name and/or e-mail address of the author. MAINTAINER details can be placed anywhere you prefer in your Dockerfile, so long as it is always post your FROM command, as they do not constitute any execution but rather a value of a definition (that is, just some additional information).

Common Constructs

```
MAINTAINER <name><email>
```

Example Dockerfile from our Joomla Image

Here, we define the author for this container and image:

```
# Add name of image author
MAINTAINER John Wooten <jwooten@cloudconsulted.com>
```

ENV

This instruction sets the environment variable in Dockerfile. An environment variable set can be used in subsequent instructions.

Common Constructs

```
ENV <key> <value>
```

The preceding code sets one environment variable `<key>` with `<value>`.

```
ENV <key1>=<value1> <key2>=<value2>
```

The preceding instruction sets two environment variables. Use the = sign between key and value of an environment variable and separate two environment key-values with space to define multiple environment variables:

```
ENV key1="env value with space"
```

Use quotes for value having spaces for environment variable.

The following are the points to remember about `ENV` instructions:

- Use single instruction to define multiple environment variables
- Environment variables are available when you create container from image
- One can review the environment variable from image using `docker inspect <image>`
- Values of environment variables can be changed at runtime by passing the `--env <key>=<value>` option to the `docker run` command

Example Dockerfile from our Joomla Image

Here, we set the environment variables for Joomla and the Docker image running without an interactive Terminal:

```
# Set the environment variables
ENV DEBIAN_FRONTEND noninteractive
ENV JOOMLA_VERSION 3.4.1
```

RUN

This instruction allows you to run commands and yield a layer. The output of the `RUN` instruction will be a layer built for image under process. Command passed to the `RUN` instruction runs on the layers built before this instruction; one needs to take care of the orders.

Common Constructs

```
RUN <command>
```

The `<command>` is executed in a shell –`/bin/sh` –`c` shell form.

```
RUN ["executable", "parameter1", "parameter2"]
```

In this particular form, you specify the `executable` and `parameters` in executable form. Ensure that you pass the absolute path of the executable in the command. This is useful for cases where the base image does not have `/bin/sh`. You can specify an executable, which could be your only executable in a base image and build the layers on top using it.

This is also useful if you do not want to use the `/bin/sh` shell. Consider this:

```
RUN ["/bin/bash", "-c", "echo True!"]
RUN <command1>;<command2>
```

Actually, this is a special form of example, where you specify multiple commands separated by `;`. The RUN instruction executes such commands together and builds a single layer for all of the commands specified.

Example Dockerfile from our Joomla Image

Here, we update the package manager and install required dependencies:

```
# Update package manager and install required dependencies
RUN apt-get update && DEBIAN_FRONTEND=noninteractive apt-get install -y \
    mysql-server \
    apache2 \
    php5 \
    php5-imap \
    php5-mcrypt \
    php5-gd \
    php5-curl \
    php5-apcu \
    php5-mysqlnd \
    supervisor
```

Note that we have purposefully written so that new packages are to be added as their own apt-get install lines, following the initial install commands.

This is done so that, should we ever need to add or remove a package, we can do so without requiring to re-install all other packages within our Dockerfile. Obviously, this provides considerable savings in build time, should the need arise.

 Docker Cache: Docker will first check against the host's image cache for any matching layers from previous builds. If found, the given build step within the Dockerfile will be skipped to utilize the previous layer, from cache. As such, it is best practice to enlist each of the Dockerfile's `apt-get -y install` commands on their own.

As we've discussed, the `RUN` command in a Dockerfile will execute any given command under the context and filesystem of the Docker container, and produce a new image layer with any resulting file system changes. We first run `apt-get update` to ensure that the repositories and the PPAs of the packages are updated. Then, in separate calls, we instruct the package manager to install MySQL, Apache, PHP, and Supervisor. The `-y` flag skips interactive confirmation.

With all of our necessary dependencies installed to run our service, we ought to tidy up a bit to give us a cleaner Docker image:

```
# Clean up any files used by apt-get
RUN apt-get clean && rm -rf /var/lib/apt/lists/* /tmp/* /var/tmp/*
```

ADD

This information is used to copy files and directories from the local filesystem or files from a remote URL into the image. The source and destination must be specified in `ADD` instructions.

Common Constructs

```
ADD   <source_file>   <destination_directory>
```

Here the path of `<source_file>` is relative to the build context. Also, the path of `<destination_directory>` could either be absolute or relative to the `WORKDIR`:

```
ADD   <file1> <file2> <file3> <destination_directory>
```

Multiple files, for example, `<file1>`, `<file2>`, and `<file3>`, are copied into `<destination_directory>`. Note that paths of these source files should be relative to the build context, as follows:

```
ADD <source_directory> <destination_directory>
```

Contents of the `<source_directory>` are copied into `<destination_directory>` along with the filesystem metadata; the directory itself is not copied:

```
ADD text_* /text_files
```

All the files starting with `text_` in the build context directory are copied in the `/text_files` directory in the container image:

```
ADD ["filename with space",...,  "<dest>"]
```

Filename with a space can be specified in quotes; one needs to use a JSON array to specify the ADD instruction in this case.

The following are the points to remember about `ADD` instructions:

- All new files and directories that are copied into the container image have UID and GID as `0`
- In cases where the source file is a remote URL, the destination file will have a permission of `600`
- All the local files referenced in the source of the `ADD` instruction should be in the build context directory or in its subdirectories
- If the local source file is a supported tar archive then it is unpacked as a directory
- If multiple source files are specified, the destination must be a directory and end with a trailing slash, `/`
- If a destination does not exist, it will be created along with all the parent directories in the path, if required

Example Dockerfile from our Joomla Image

Here, we download `joomla` into the Apache web root:

```
# Download joomla and put it default apache web root
ADD
https://github.com/joomla/joomla-cms/releases/download/$JOOMLA_VERSION/Joomla_$JOOMLA_VERSION-Stable-Full_Package.tar.gz /tmp/joomla/
RUN tar -zxvf /tmp/joomla/Joomla_$JOOMLA_VERSION-Stable-Full_Package.tar.gz -C /tmp/joomla/
RUN rm -rf /var/www/html/*
```

```
RUN cp -r /tmp/joomla/* /var/www/html/

# Put default htaccess in place
RUN mv /var/www/html/htaccess.txt /var/www/html/.htaccess

RUN chown -R www-data:www-data /var/www

# Expose HTTP and MySQL
EXPOSE 80 3306
```

COPY

The COPY command specifies that a file, located at the input path, should be copied from the same directory as the Dockerfile to the output path inside the container.

CMD

The CMD instruction has three forms–a shell form, as default parameters to ENTRYPOINT and the preferred executable form. The main purpose of a CMD is to provide defaults for an executing container. These defaults can either include or omit an executable, the latter of which must specify an ENTRYPOINT instruction as well. If the user specifies arguments to Docker run, then they will override the default specified in CMD. If you would like your container to run the same executable every time, then you should consider using ENTRYPOINT in combination with CMD.

The following are the points to remember:

- Do not to confuse CMD with RUN—RUN will actually execute the command and commit the result, whereas CMD does not execute commands during a build, but instead specifies the intended command for the image
- A Dockerfile can only execute one CMD; if you enlist more than one, only the last CMD will be executed

Example Dockerfile from our Joomla Image

Here, we set up Apache for it to start:

```
# Use supervisord to start apache / mysql
COPY supervisord.conf /etc/supervisor/conf.d/supervisord.conf
CMD ["/usr/bin/supervisord", "-n"]
```

The following is the content of our completed Joomla Dockerfile:

```
FROM ubuntu
MAINTAINER John Wooten <jwooten@cloudconsulted.com>

ENV DEBIAN_FRONTEND noninteractive
ENV JOOMLA_VERSION 3.4.1

RUN apt-get update && DEBIAN_FRONTEND=noninteractive apt-get install -y \
    mysql-server \
    apache2 \
    php5 \
    php5-imap \
    php5-mcrypt \
    php5-gd \
    php5-curl \
    php5-apcu \
    php5-mysqlnd \
    supervisor

# Clean up any files used by apt-get
RUN apt-get clean && rm -rf /var/lib/apt/lists/* /tmp/* /var/tmp/*

# Download joomla and put it default apache web root
ADD
https://github.com/joomla/joomla-cms/releases/download/$JOOMLA_VERSION/Joom
la_$JOOMLA_VERSION-Stable-Full_Package.tar.gz /tmp/joomla/
RUN tar -zxvf /tmp/joomla/Joomla_$JOOMLA_VERSION-Stable-Full_Package.tar.gz
-C /tmp/joomla/
RUN rm -rf /var/www/html/*
RUN cp -r /tmp/joomla/* /var/www/html/

# Put default htaccess in place
RUN mv /var/www/html/htaccess.txt /var/www/html/.htaccess

RUN chown -R www-data:www-data /var/www

# Expose HTTP and MySQL
EXPOSE 80 3306

# Use supervisord to start apache / mysql
COPY supervisord.conf /etc/supervisor/conf.d/supervisord.conf
CMD ["/usr/bin/supervisord", "-n"]
```

Other common Dockerfile commands are as follows:
ENTRYPOINT

An `ENTRYPOINT` allows you to configure a container that will run as an executable. From Docker's documentation, we will use the provided example; the following will start `nginx` with its default content, listening on port `80`:

```
docker run -i -t --rm -p 80:80 nginx
```

Command-line arguments to `docker run <image>` will be appended after all elements in an executable form `ENTRYPOINT`, and will override all elements specified using `CMD`. This allows arguments to be passed to the entry point, that is, `docker run <image> -d` will pass the `-d` argument to the entry point. You can override the `ENTRYPOINT` instruction using the `docker run --entrypoint` flag.

LABEL

This instruction specifies the metadata for the image. This image metadata can later be inspected using the `docker inspect <image>` command. The idea here is to add information about the image in image metadata for easy retrieval. In order to get the metadata from the image, one does not need to create a container from the image (or mount the image on local filesystem), Docker associates metdata data with every Docker image, and it has a predefined structure for it; using `LABEL`, one can add additional associated metadata describing the image.

The label for the image is a key-value pair. Following are examples of using `LABEL` in a Dockerfile:

```
LABEL <key>=<value>   <key>=<value>   <key>=<value>
```

This instruction will add three labels to the image. Also, note that it will create one new layer as all the labels are added in a single `LABEL` instruction:

```
LABEL  "key"="value with spaces"
```

Use quotes in labels if the label value has spaces:

```
LABEL LongDescription="This label value extends over new \
line."
```

If the value of the label is long, use backslash to extend the label value to a new line.

```
LABEL key1=value1
LABEL key2=value2
```

Multiple labels for an image can be defined by separating them by **End Of Line** (EOL). Note that, in this case, there will be two image layers created for two different LABEL instructions.

Notes about LABEL instructions:

- Labels are collated together as described in Dockerfile and those from the base image specified in the FROM instruction
- If key in labels are repeated, later one will override the earlier defined key's value.
- Try specifying all the labels in a single LABEL instruction to produce an efficient image, thus avoiding unnecessary image layer count
- To view the labels for a built image, use the docker inspect <image> command

WORKDIR

This instruction is used to set the working directory for subsequent RUN, ADD, COPY, CMD, and ENTRYPOINT instructions in Dockerfile.

Define a work directory in Dockerfile, all subsequent relative paths referenced inside the container will be relative to the specified work directory.

The following are examples of using the WORKDIR instruction:

```
WORKDIR /opt/myapp
```

The preceding instruction specifies /opt/myapp as the working directory for subsequent instructions, as follows:

```
WORKDIR /opt/
WORKDIR myapp
RUN pwd
```

The preceding instruction defines the work directory twice. Note that the second WORKDIR will be relative to the first WORKDIR. The result of the pwd command will be /opt/myapp:

```
ENV SOURCEDIR /opt/src
WORKDIR $SOURCEDIR/myapp
```

Work directory can resolve the environment variables defined earlier. In this example, the WORKDIR instruction can evaluate the SOURCEDIR environment variable and the resultant working directory will be /opt/src/myapp.

USER

This sets the user for running any subsequent `RUN`, `CMD`, and `ENTRYPOINT` instructions. This also sets the user when a container is created and run from the image.

The following instruction sets the user `myappuser` for the image and container:

```
USER myappuser
```

Notes about `USER` instructions:

- One can override the user using `--user=name|uid[:<group|gid>]` in the `docker run` command for container

Image testing and debugging

While we can applaud the benefits of containers, troubleshooting and effectively monitoring them currently present some complexity. Since by design, containers run in isolation, their resulting environment can be cloudy. Effective troubleshooting has generally required shell entry into the container itself, coupled with the complications of installing additional Linux tools to merely peruse information that is twice as hard to investigate.

Typically, available tools, methods, and approaches for meaningful troubleshooting of our containers and images has required installing additional packages in every container. This results in the following:

- Requirements for connecting or attaching directly to the container, which is not always a piddling matter
- Limitations on inspection of a single container at a time

Compounding these difficulties, adding unnecessary bloat to our containers with these tools is something we originally attempted to avoid in our planning; minimalism is one of the advantages we looked for in using containers in the first place. Let's take a look then at how we can reasonably glean useful information on our container images with some basic commands, as well as investigate emergent applications that allow us to monitor and troubleshoot containers from the outside.

Docker details for troubleshooting

Now that you have your image (regardless of building method) with Docker running, let's do some testing to make sure that all is copacetic with our build. While these may seem routine and mundane, it is a good practice to run any or all of the following as a *top-down* approach to troubleshooting.

The first two commands here are ridiculously simple and seemingly too generic, but will provide base-level detail with which to begin any downstream troubleshooting efforts–$ `docker version` and $ `docker info`.

Docker version

Let's ensure that we firstly recognize what version of Docker, Go, and Git we are running:

```
$ sudo docker version
```

Docker info

Additionally, we should understand our host operating system and kernel version, as well as storage, execution, and logging drivers. Knowing these things can help us troubleshoot from our *top-down* perspective:

```
$ sudo docker info
```

A troubleshooting note for Debian/Ubuntu

From a $ `sudo docker info` command, you may receive one or both of the following warnings:

```
WARNING: No memory limit support
WARNING: No swap limit support
```

You will need to add the following command-line parameters to the kernel in order to enable memory and swap accounting:

```
cgroup_enable=memory swapaccount=1
```

For these Debian or Ubuntu systems, if you use the default GRUB bootloader, those parameters can be added by editing /etc/default/grub and extending GRUB_CMDLINE_LINUX. Locate the following line:

```
GRUB_CMDLINE_LINUX=""
```

Then, replace it with the following one:

```
GRUB_CMDLINE_LINUX="cgroup_enable=memory swapaccount=1"
```

Then, run update-grub and reboot the host machine.

Listing installed Docker images

We also need to ensure that the container instance has actually installed your image locally. SSH into the docker host and execute the docker images command. You should see your docker image listed, as follows:

```
$ sudo docker images
```

What if my image does not appear? Check the agent logs and make sure that your container instance is able to contact your docker registry by curling the registry and printing out the available tags:

```
curl [need to add in path to registry!]
```

 What $ sudo docker images tells us: Our container image was successfully installed on the host.

Manually crank your Docker image

Now that we know our image is installed on the host, we need to know whether it is accessible to the Docker daemon. An easy way to test to make certain your image can be run on the container instance is by attempting to run your image from the command line. There is an added benefit here: we will now have the opportunity to additionally inspect application logs for further troubleshooting.

Let's take a look at the following example:

```
$ sudo docker run -it [need to add in path to registry/latest bin!]
```

What $ sudo docker run <imagename> tells us: Our container image is accessible from the docker daemon and also provides accessible output logs for further troubleshooting.

What if my image does not run? Check for any running containers. If the intended container isn't running on the host, there may be issues preventing it from starting:

```
$ sudo docker ps
```

When a container fails to start, it does not log anything. Output of logs for container start processes are located in `/var/log/containers` on the host. Here, you will find files following the naming convention of `<service>_start_errors.log`. Within these logs, you will find any output generated by our RUN command, and are a recommended starting point in troubleshooting as to why your container failed to start.

 TIP: Logspout (`https://github.com/gliderlabs/logspout`) is a log router for Docker containers that runs inside Docker. Logsprout attaches to all containers on a host, then routes their logs wherever you desire.

While we can also peruse the `/var/log/messages` output in our attempts to troubleshoot, there are a few other avenues we can persue, albeit a little more labor intensive.

Examining the filesystem state from cache

As we've discussed, after each successful RUN command in our Dockerfiles, Docker caches the entire filesytem state. We can exploit this cache to examine the latest state prior to the failed RUN command.

To accomplish the task:

- Access the Dockerfile and comment out the failing RUN command, in addition to any and subsequent RUN commands
- Re-save the Dockerfile
- Re-execute $ `sudo docker build` **and** $ `sudo docker run`

Image layer IDs as debug containers

Every time Docker successfully executes a RUN command from a Dockerfile, a new layer in the image filesystem is committed. Conveniently, you can use those layers IDs as images to start a new container.

Consider the following Dockerfile as an example:

```
FROM centos
RUN echo 'trouble' > /tmp/trouble.txt
RUN echo 'shoot' >> /tmp/shoot.txt
```

If we then build from this Dockerfile:

```
$ docker build -force-rm -t so26220957 .
```

We would get output similar to the following:

```
Sending build context to Docker daemon 3.584 kB
Sending build context to Docker daemon
Step 0 : FROM ubuntu
  ---> b750fe79269d
Step 1 : RUN echo 'trouble' > /tmp/trouble.txt
  ---> Running in d37d756f6e55
  ---> de1d48805de2
Removing intermediate container d37d756f6e55
Step 2 : RUN echo 'bar' >> /tmp/shoot.txt
Removing intermediate container a180fdacd268
Successfully built 40fd00ee38e1
```

We can then use the preceding image layer IDs to start new containers from b750fe79269d, de1d48805de2, and 40fd00ee38e1:

```
$ docker run -rm b750fe79269d cat /tmp/trouble.txt
cat: /tmp/trouble.txt No such file or directory
$ docker run -rm de1d48805de2 cat /tmp/trouble.txt
trouble
$ docker run -rm 40fd00ee38e1 cat /tmp/trouble.txt
trouble
shoot
```

 We employ --rm to remove all the debug containers since there is no reason to have them around postruns.

What happens if my container build fails? Since no image is created on a failed build, we'd have no hash of the container with which to ID. Instead, we can note the ID of the preceding layer and run a container with a shell of that ID:

```
$ sudo docker run --rm -it <id_last_working_layer> bash -il
```

Once inside the container, execute the failing command in attempt to reproduce the issue, fix the command and test, and finally update the Dockerfile with the fixed command.

You may also want to start a shell and explore the filesystem, try out commands, and others:

```
$ docker run -rm -it de1d48805de2 bash -il
root@ecd3ab97cad4:/# ls -l /tmp
total 4
-rw-r-r-- 1 root root 4 Jul 3 12:14 trouble.txt
root@ecd3ab97cad4:/# cat /tmp/trouble.txt
trouble
root@ecd3ab97cad4:/#
```

Additional example

One final example is to comment out of the following Dockerfile, including the offending line. We are then able to run the container and docker commands manually and look into the logs in the normal way. In this example Dockerfile:

```
RUN trouble
RUN shoot
RUN debug
```

Also, the failure is at shoot, then comment out as follows:

```
RUN trouble
# RUN shoot
# RUN debug
```

Then, build and run:

```
$ docker build -t trouble .
$ docker run -it trouble bash
container# shoot
...grep logs...
```

Checking failed container processes

Even if your container successfully runs from the command line, it would prove beneficial to inspect for any failed container processes, for containers that are no longer running, and checking our container configuration.

Run the following command to check for failed or no-longer running containers and note the CONTAINER ID to inspect a given container's configuration:

```
$ sudo docker ps -a
```

Note the **STATUS** of the containers. Should any of your containers, **STATUS** show exit codes other than 0, there could be issues with the container's configuration. By way of an example, a bad command would result in an exit code of 127. With this information, you can troubleshoot the task definition CMD field to debug.

Although somewhat limited, we can further inspect a container for additional troubleshooting details:

```
$ sudo docker inspect <containerId>
```

Finally, let's also analyze the container's application logs. Error messages for container start failures are output here:

```
$ sudo docker logs <containerId>
```

Other potentially useful resources

$ sudo docker top gives us a list of processes running inside a container.

$ sudo docker htop can be utilized when you need a little more detail than provided by top in a convenient, cursor-controlled inferface. htop starts faster than top, you can scroll the list vertically and horizontally to see all processes and complete command lines, and you do not need to type the process number to kill a process or the priority value to recieve a process.

By the time this book goes to print, it is likely that the mechanisms for troubleshooting containers and images will have dramatically improved. Much focus is being given by the Docker community toward *baked-in* reporting and monitoring solutions, in addition to market forces that will certainly bring additional options to bear.

Using sysdig to debug

As with any newer technology, some of the initial complexities inherent with them are debugged in time, and newer tools and applications are developed to enhance their use. As we've discussed, containers certainly fit into this category at this time. While we have witnessed improvements in availability of official, standardized images within the Docker Registry, we are also now seeing emergent tools that help us to effectively manage, monitor, and troubleshoot our containers.

Sysdig provides application monitoring for containers [Image Copyright © 2014 Draios, Inc.]

Sysdig (`http://www.sysdig.org/`) is one such tool. As an *au courant* application for system-level exploration and troubleshooting visibility into containerized environments, the beauty of `sysdig` is that we are able to access container data from the outside (even though `sysdig` can actually also be installed inside a container). From a top level, what `sysdig` brings to our container management is this:

- Ability to access and review processes (inclusive of internal and external PIDs) in each container
- Ability to drill-down into specific containers
- Ability to easily filter sets of containers for process review and analysis

Sysdig provides data on CPU usage, I/O, logs, networking, performance, security, and system state. To repeat, this is all accomplishable from the outside, without a need to install anything into our containers.

We will make continued and valuable use of `sysdig` going forward in this book to monitor and troubleshoot specific processes related to our containers, but for now we will provide just a few examples toward troubleshooting our basic container processes and logs.

Let's dig into `sysdig` by getting it installed on our host to show off what it can do for us and our containers!

Single step installation

Installation of `sysdig` can be accomplished in a single step by executing the following command as root or with `sudo`:

```
curl -s https://s3.amazonaws.com/download.draios.com/stable/install-sysdig
| sudo bash
```

 NOTE: `sysdig` is currently included natively in the latest Debian and Ubuntu versions; however, it is recommended to update/run installation for the latest packages.

Advanced installation

According to the `sysdig` wiki, the advanced installation method may be useful for scripted deployments or containerized environments. It is also easy; the advanced installation method is enlisted for RHEL and Debian systems.

What are chisels?

To get started with `sysdig`, we should understand some of its parlance, specifically **chisels**. In `sysdig`, chisels are little scripts (written in Lua) that analyze the `sysdig` event stream to perform useful actions. Events are efficiently brought to user level, enriched with context, and then scripts can be applied to them. Chisels work well on live systems, but can also be used with trace files for offline analysis. You can run as many chisels as you'd like, all at the same time. For example:

`topcontainers_error` chisel will show us the top containers by number of errors.

For a list of sysdig chisels:

`$ sysdig -cl` (use the `-i` flag to get detailed information about a specific chisel)

Single container processes analysis

Using the example of a `topprocs_cpu` chisel, we can apply a filter:

```
$ sudo sysdig -pc -c topprocs_cpu container.name=zany_torvalds
```

These are the example results:

```
CPU%            Process        container.name
-------------------------------------------------
02.49%          bash           zany_torvalds
37.06%          curl           zany_torvalds
0.82%           sleep          zany_torvalds
```

Unlike using `$ sudo docker top` (and similar), we can determine exactly which containers we want to see processes for; for example, the following example shows us processes from only the `wordpress` containers:

```
$ sudo sysdig -pc -c topprocs_cpu container.name contains wordpress
```

```
CPU%            Process        container.name
-------------------------------------------------
5.38%           apache2        wordpress3
4.37%           apache2        wordpress2
6.89%           apache2        wordpress4
7.96%           apache2        wordpress1
```

Other Useful Sysdig Chisels & Syntax

- `topprocs_cpu` shows top processes by CPU usage
- `topcontainers_file` shows top containers by R+W disk bytes
- `topcontainers_net` shows top containers by network I/O
- `lscontainers` will list the running containers
- `$ sudo sysdig -pc -cspy_logs` analyzes all logs per screen
- `$ sudo sysdig -pc -cspy_logs container.name=zany_torvalds` prints logs for the container `zany_torvalds`

Troubleshooting – an open community awaits you

In general, most issues you may face have likely been experienced by others, somewhere and sometime before. The Docker and open source communities, IRC channels and various search engines, can provide resulting information that is highly accessible and likely to provide you with answers to situations, and conditions, that perplex. Make good use of the open source community (specifically, the Docker community) in getting the answers you are looking for. As with any emergent technology, in the beginning, we are all somewhat learning together!

Automated image building

There are many ways we can go about automating our processes for building container images; too many to reasonably provide a full disclosure of approaches within a single book. In later chapters of this book, we will delve more deeply into a range of automation options and tools. In this particular instance, we are only speaking of automation using our Dockerfile. We have already discussed in general that Dockerfiles can be used in automating our image building, so let's take a more dedicated look into Dockerfile automation specifically.

Unit tested deployments

During the build process, Docker allows us to run any command. Let's take advantage of this to enable unit tests while building our image. These unit tests can help to identify problems in our production image before we push them to staging or deployment, and will at least partially verify the image functions the way we intend and expect. If the unit tests run successfully, we have a degree of confidence that we have a valid runtime environment for our service. This also means that should the tests fail, our build will fail, effectively keeping a nonworking image out of its production.

Using our `cloudconsulted/joomla` repository image from prior, we will set up a sample workflow for automated builds, with testing. **PHPUnit** is what we will use since it is officially used by the Joomla! project's development teams, as it can conveniently run unit tests against our entire stack–the Joomla code, Apache, MySQL, and PHP.

Drop in to your Dockerfile directory for `cloudconsulted/joomla` (in our case, `dockerbuilder`) and update it as follows.

Install PHPUnit executing the following commands:

```
[# install composer to a specific directory
curl -sS https://getcomposer.org/installer | php -- --install-dir=bin
# use composer to install phpunit
composer global require "phpunit/phpunit=4.1.*"]
```

PHPUnit can also be installed executing the following commands:

```
[# install phpunit
wget https://phar.phpunit.de/phpunit.phar
chmod +x phpunit.phar
mv phpunit.phar /usr/local/bin/phpunit
# might also need to put the phpunit executable placed here? test this:
cp /usr/local/bin/phpunit /usr/bin/phpunit]
```

Now, let's run our unit tests with `phpunit`:

```
# discover and run any tests within the source code
RUN phpunit
```

We also need to make sure that we COPY our unit tests to the assets inside our image:

```
# copy unit tests to assets
COPY test /root/test
```

Lastly, let's do some house cleaning. To ensure that our production code cannot rely (accidentally or otherwise) on the test code, once the unit tests complete we should delete those test files:

```
# clean up test files
RUN rm -rf test
```

Our total updates to the Dockerfile included:

```
wget https://phar.phpunit.de/phpunit.phar
chmod +x phpunit.phar
mv phpunit.phar /usr/local/bin/phpunit

RUN phpunit
COPY test /root/test
RUN rm -rf test
```

Now, we have a scripted Dockerfile that, each and every time we build this image, will fully test our Joomla code, Apache, MySQL, and PHP dependencies as a literal part of the build process. The results are a tested, reproducible production environment!

Automating tested deployments

With our heightened confidence in producing workable images for deployment, this build process still requires a developer or DevOps engineer to rebuild the image before every production push. Instead, we will rely on automated builds from our Docker and GitHub repositories.

Our GitHub and Docker Hub repositories will serve to automate our builds. By maintaining our Dockerfiles, dependencies, related scripts, and so on on GitHub, any pushes or commits to update files on the repository will automatically force an updating push to the synced Docker Hub repository. Our production images for pull on Docker Hub are automatically updated with any new build information.

Docker Clouds is one of the latest offerings to complete the app life cycle, it provides a hosted registry service with build and testing facilities. Docker Cloud expands on the feature of Tutum and brings a tighter integration with Docker Hub. With the help of a Docker Cloud system, admins can deploy and scale applications in the cloud with just a few clicks. Continuous deliver the code integrated and automated with build, test and deployment workflows. It also provides visibility across the containers of the entire infrastructure and accesses the programmatic RESTful APIs for a developer-friendly CLI tool. Thus, Docker Cloud can be used for automating the build process and test deployments.

The following are the important features of Docker Cloud:

- Allows the building of Docker images and also linking cloud repositories to a source code in order to ease the process of image building
- It allows linking your infrastructure and cloud services to provision new nodes automatically
- Once the image has been built, it can be used to deploy services and can be linked with Docker Cloud's collection of services and microservices
- Swarm management in beta mode is available for creating swarm within Docker Clouds or registering the existing swarms to Docker Clouds using Docker ID

Summary

Docker and Dockerfiles provide repeatable processes across the application development cycle, providing a distinctive facility for both developers and DevOps engineers–production-ready deployments, infused with the confidence of tested images and the ease of automation. This provides a high level of empowerment to those needing it most, and results in the continuous delivery of tested and production-ready image building that we can fully automate, extended as far out as, and across, our clouds.

In this chapter, we learned that a mission-critical task in a production-ready application containerization is image building. The building of base and layered images and avoiding areas for troubleshooting are the primary topics we covered. In building our base images, we saw that the Docker Registry provides ample and validated images that we can freely use for repeatable processes. We also canvassed building images manually, from scratch. Moving forward, we explored building layered images with a Dockerfile and enlisted the Dockerfile commands in detail. Finally, an example workflow illustrated automated image building with baked-in testing of images and containers. Throughout, we highlighted the ways and means for troubleshooting areas and options.

Building succinct Docker images for your application container is vitally crucial for your application's functionality and maintainability. Now that we have learned about building base and layered images and basic ways to troubleshoot them, we will look foward to building real application images. In our next chapter, we will learn about planning and building multiple-tier applications with a proper set of images.

4
Devising Microservices and N-Tier Applications

Let's expand on what we saw and learned in the last chapter about the more advanced development and deployment of microservices and N-tier applications. This chapter will address the underlying architectures for these design approaches as well as resolve typical issues faced while building these types of applications. We will cover the following topics in the chapter:

- Monolithic architectural pattern
- N-tier application architecture
- Building, testing, and automating N-tier applications
- Microservices architectural pattern
- Building, testing, and automating microservices
- Decoupling multi-tier applications into multiple images
- Making different tiers of applications work

Nowadays, modern software built as services are giving rise to a shift in how applications are designed. Instead of using web frameworks to invoke services and produce web pages, applications today are built by consuming and producing APIs. Much has changed in the development and deployment of business applications, some of it dramatically and some of it either by revision or extension from the past design approaches, depending upon your viewpoint. Several architectural design approaches exist, and they are distinguishable by applications built for enterprise versus web versus Cloud.

Development trends, over the last few years in particular, are awash with terms such as **microservices architecture** (**MSA**), applicable to a particular way of application design and development as suites of independently deployable services. The meteoric rise of the microservices architectural style is clearly an irrefutable force in today's development for deployment; there has been a considerable shift away from monolithic architecture and toward N-tier applications and microservices, but just how much of this is hype and how much of this can be honed?

Hype or hubris

Before we begin diving deeply into troubleshooting, we ought to provide a basic contextual overview of modern applications and both the N-tier and microservices architectural styles. Knowing both the advantages and limitations of these architectural styles will help us plan for potential troubleshooting areas, and how we can avoid them. Containers are ideally suited for both of these architectural approaches, and we will discuss each one separately to give their proper due.

Within all the noise, we sometimes forget that to deploy systems across these domains, one still has to create services and compose multiple services in working distributed applications. Here, it is important to understand the modern meaning of the term application. Applications are now primarily constructed as asynchronous message flows or synchronous request calls (if not both) that serve in forming collections of components or services allied by these connections. Participating services are highly distributive across variant machines and diverse Clouds (private, public, and hybrid).

As for architectural styles, we shan't bother ourselves with too much comparison or engage in overly detailed discussions on what microservices actually are and whether they are any different from **Service-Oriented Architecture** (**SOA**)–there is certainly plenty of forum and related debate elsewhere for your choosing. With design principles rooted at least as far back as Unix, we will proffer no authoritative viewpoints in this book that the current microservices trend is either conceptually singular or entirely ingenious. Instead, we will put forward the major considerations for implementing this architectural approach and the benefits to be gained for modern applications.

Use case still drives and dictates architectural approaches (or, in my opinion, should), and as such there is value in making some degree of comparative analysis among all predominant architectural styles: **monolithic**, **N-tier**, and **microservices**.

Monolithic architecture

Monoliths are essentially one deployment unit housing all services and dependencies, making them easy to develop, easy to test, relatively easy to deploy and, initially, easy to scale. However, this style does not meet the requisite needs for most modern enterprise applications (N-tier) and web development at scale, and certainly not (microservices) applications being deployed to the Cloud. Change cycles are tightly coupled–any changes made, even to the smallest parts of an application, require wholesale rebuilds and redeployments for the entire monolith. As the monolith matures, any attempts at scaling require scaling of the entire application rather than the individual parts, which specifically require greater resources, becoming altogether nightmarish, if not improbable. At this point, a monolithic application has become overly complex, weighted with vast lines of code that is ever-increasingly difficult to decipher, such that business-critical items like bug fixes or implementing new features become too much of a time drain to ever attempt. As the code base becomes unintelligible, it is only reasonable to expect any changes made likely to be done incorrectly. The burgeoning size of the application not only slows development, it impedes continuous development altogether; to update any part of a monolith, the entire app must be redeployed.

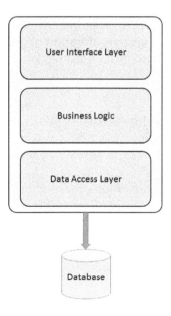

Monolithic architectural pattern

Other problems with monoliths abound, resources cannot be catered to better meet needs, for example, CPU or memory requirements. Since all modules are running the same processes, any bug can potentially bring the entire process to a halt. Lastly, it becomes much more difficult to adopt newer frameworks or languages, creating a huge barrier to adopt new technologies–you are likely stuck with whatever technology choices you made at the beginning of the project. Needless to say, your needs may have changed rather dramatically since the beginning. Using obsolete, unproductive technology makes keeping and bringing in new talent more difficult. The application has now become very difficult to scale and is unreliable, making agile development and delivery of applications impossible. The initial ease and simplicity of a monolith quickly become its own Achilles heel.

As these monolithic architectures are basically one deployment unit that does everything–N-tier and microservices architectures have arisen to address the specialized service needs of modernized applications, primarily Cloud and mobile-based.

N-tier application architecture

In order to understand N-tier applications and their potential for decoupling into microservices, we will hold its comparison against the monolithic style since both the development of N-tier applications and proliferation of microservices exist to address many of the problems found in the outdated conditions we've found resulting from the approach of monolithic architectures.

The N-tier application architecture, also referred to as **distributed applications** or **multi-tier**, proffers a model in which developers can create flexible and reusable applications. As the application is segregated into tiers, developers are empowered by the option of modifying or adding a specific tier or layer instead of requiring a rework of the entire application as would be necessary under monolithic. A multi-tier application is any application developed and distributed among more than one layer. It logically separates the different application-specific and operational layers. The number of layers varies by business and application requirements, but three-tier is the most commonly used architecture. Multi-tier applications are used to divide enterprise applications into two or more components that may be separately developed, tested, and deployed.

N-tier applications are essentially SOA that attempt to address some of the issues with antiquated monolothic design architecture. As we have seen in the previous chapters, Docker containers are a perfect match for N-tier application development.

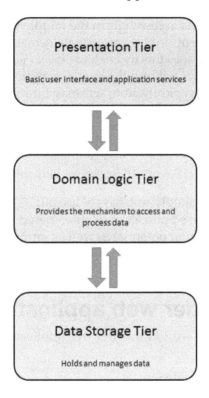

N-tier application architecture

A common N-tier application consists of three layers: a **PRESENTATION TIER** (providing basic user interface and application services access), a **DOMAIN LOGIC TIER** (providing the mechanism used to access and process data), and a **DATA STORAGE TIER** (which holds and manages data that is at rest).

While the concepts of layer and tier are often used interchangeably, a fairly common point of view is that there is actually a difference. This view holds that a *layer* is a logical structuring mechanism for the elements that make up the software solution, while a *tier* is a physical structuring mechanism for the system infrastructure. Unless otherwise specifically noted in our book, we will use tier and layer interchangeably.

The easiest way to separate the various tiers in an N-tier application is to create discrete projects for each tier that you want to include in your application. For example, the presentation tier might be a Windows forms application, whereas the data access logic might be a class library located in the middle tier. Additionally, the presentation layer might communicate with the data access logic in the middle tier through a service. Separating application components into separate tiers increases the maintainability and scalability of the application. It does this by enabling easier adoption of new technologies that can be applied to a single tier without the requirement to redesign the whole solution. In addition, N-tier applications typically store sensitive information in the middle tier, which maintains isolation from the presentation tier.

Probably the most common example of N-tier app development is websites; an example of this can be seen in the `cloudconsulted/joomla` image we used in the last chapter, where Joomla, Apache, MySQL, and PHP were all *layered* as tiers into a single container.

It will be easy enough for us to simply recursively use our `cloudconsulted/joomla` image (from earlier) here, but let's build a classic three-tiered web application to expose ourselves to some other application potential as well as introduce another unit test tool for our development teams.

Building a three-tier web application

Let's develop and deploy a real-world three-tier web application with the help of the following containers:

NGINX > Ruby on Rails > PostgreSQL:

The NGINX Docker container (Dockerfile), as follows:

```
## AngularJS Container build
FROM nginx:latest

# Download packages
RUN apt-get update
RUN apt-get install -y curl    \
                 git    \
                 ruby \
                 ruby-dev \
                 build-essential

# Copy angular files
COPY . /usr/share/nginx

# Installation
```

```
RUN curl -sL https://deb.nodesource.com/setup | bash -
RUN apt-get install -y nodejs \
                    rubygems
RUN apt-get clean
WORKDIR /usr/share/nginx
RUN npm install npm -g
RUN npm install -g bower
RUN npm install  -g grunt-cli
RUN gem install sass
RUN gem install compass
RUN npm cache clean
RUN npm install
RUN bower -allow-root install -g

# Building
RUN grunt build

# Open port and start nginx
EXPOSE 80
CMD ["/usr/sbin/nginx", "-g", "daemon off;"]
```

The Ruby on Rails Docker container (Dockerfile), as shown:

```
## Ruby-on-Rails Container build
FROM rails:onbuild

# Create and migrate DB
RUN bundle exec rake db:create
RUN bundle exec rake db:migrate

# Start rails server
CMD ["bundle", "exec", "rails", "server", "-b", "0.0.0.0"]
```

The PostgreSQL Docker container, as illustrated:

```
## PostgreSQL Containers build
# cloudconsulted/postgres is a Postgres setup that accepts remote
connections from Docker IP (172.17.0.1/16).  We can therefore make use of
this image directory so there is no need to create a new Docker file here.
```

The preceding Dockerfiles can be used to deploy a three-tier web application and help us get started with microservices.

Microservices architecture

To begin explaining the microservice architectural style, it will prove beneficial to again compare to the monolithic, as we did with N-tier. As you may recall, a monolithic application is constructed as a single unit. Also, recall that monolithic enterprise applications are often built around three primary tiers: a client-side user interface (comprising of HTML pages and JavaScript running in a browser on the user's machine), a database (comprising of many tables inserted into a common, and usually relational, database management system), and a server-side application (which handles HTTP requests, executes domain logic, retrieves and updates data from the database, and selects and populates HTML views to be sent to the browser). This classic version of a monolithic enterprise application is a single, logical executable. Any changes to the system involve building and deploying a new version of the server-side application, and changes in the underlying technology are likely not prudent.

The path to modernity

Microservices represent the convergence of the modern Cloud and modern application development, structured around the following:

- Componentized services
- Organization around business capabilities
- Products, not projects
- Smart endpoints and dumb pipes
- Decentralized governance and data management
- Infrastructure automation

Here, monolithic typically focuses on **enterprise service bus** (**ESB**) used to integrate monolithic applications, modern application design is API driven. These modern applications embrace APIs on all sides: on the frontend for connecting to rich clients, the backend for integrating with internal systems, and on the sides to allow other applications access to their internal data and processes. Rather than leveraging the more complicated traditional enterprise mechanisms, many developers are finding that the same lightweight API services that have proven to be resilient, scalable, and agile for frontend, backend, and application-to-application scenarios can also be leveraged for application assembly. What is also compelling is that containers, and especially so within a microservices architectural approach, alleviate the perennial issue of developers being blocked out of architectural decisions while still realizing the benefits of repeatability. The use of preapproved container configurations.

Microservices architectural pattern

Here, we illustrate that instead of a single, monstrous monolithic application, we have split the application into smaller, interconnected services (that is, microservices) that implement each functional area of the application. This allows us to deploy directly to address the needs of specialized use cases or specific devices or users /or/ the microservices approach, in a nutshell, dictates that instead of having one giant code base that all developers touch, which often becomes perilous to manage, there are numerous smaller code bases managed by small and agile teams. The only dependency these code bases have on one another is their APIs:

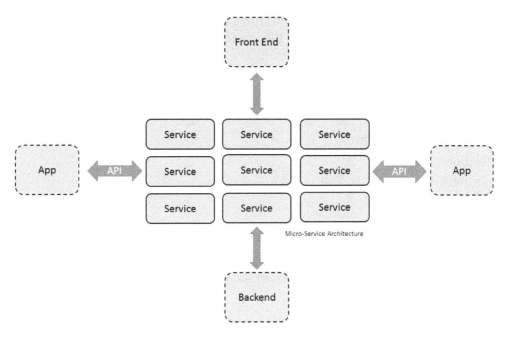

Microservices architectural pattern

 A common discussion around microservices is debate over whether this is just SOA. Some validity exists on this point as the microservice style does share some of the advocacies of SOA. In reality, SOA means a host of many different things. As such, we submit and will attempt to show that while shared similarities do exist, SOA remains significantly different from the microservices architectural style as presented herein.

Common characteristics of microservices

While we will not attempt a formal definition of the microservices architectural style, there are some common characteristics we can certainly use to identify it. Microservices are generally designed around business capabilities and priorities and include multiple component services that can be automated for deployment independently without compromising the application, intelligence endpoints, and decentralized control of languages and data.

To provide some basis then, if not common ground, to follow is an outline that can be seen as the common characteristics for architectures that fit the *microservices* label. It should be understood that not all microservice architectures will exhibit all characteristics at all times. Since we do, however, have expectations that most microservice architectures will exhibit most of these characteristics, let's list them:

- Independent
- Stateless
- Asynchronous
- Single responsibility
- Loosely coupled
- Interchangeable

Advantages of microservices

The common characteristics of microservices we just listed also serve to itemize their advantages. Without meaning to belabor the issue over too much redundancy, let's at least canvass the main advantage points:

- **Microservices enforce a level of modularity**: This is extremely difficult to accomplish in practice with a monolithic architecture. The microservices advantage is that individual services are much faster to develop, much easier to understand, and much easier to maintain.

- **Microservices enable each service to be developed independently**: This is done by teams specifically focused on that service. The microservices advantage is empowering developers with the freedom to choose whatever technology is best suited or makes better sense, so long as that service honors the API contract. By default, this also means that developers are no longer trapped with potentially obsolete technologies from a project's beginning, or when starting a new project. Not only does an option exist to employ the current technology, but with a relatively small service size it is also now feasible to rewrite older services using a more relevant and reliable technology.

- **Microservices enable each service to be deployed continuously**: Developers needn't coordinate the deployment of changes that are localized to their service. The microservices advantage is in continuous deployment–deployment takes place as soon as changes are successfully tested.

- **Microservices enable each service to be scaled independently**: You need to deploy only the instances of each service necessary to satisfy the capacity and availability constraints. Additionally, we can also succinctly match the hardware to fulfill a service's resource requirements (for example, compute or memory optimized hardware for CPU and memory-intensive services). The microservices advantage is in not only matching capacity and availability, but leveraging user-specific hardware optimized for a service.

All of these advantages are extremely advantageous, but for the next bit let's elaborate on the point of scalability. As we've seen with monolithic architectures, while easy to initialize scaling, it is certainly deficient in executing it over time; bottlenecks abound and, eventually, it's approach to scaling is vastly untenable. Fortunately, microservices as an architectural style supremely excels at scaling. A quintessential book, *THE ART OF SCALABILITY* (http://theartofscalability.com/) illustrates a highly useful, three-dimensional model of scalability in a *scale cube* (http://microservices.io/articles/scal ecube.html).

Microservices at scalability

In the provided model, along the X-axis scaling (that is, Monolothic) we can see the common horizontal duplication approach, scaling an application by running multiple, cloned copies of that application behind a load balancer. This results in improved application capacity and availability.

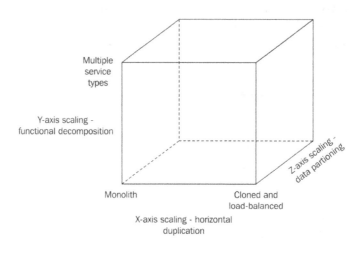

Microservices at scalability

Moving along the Z-axis for scaling (that is N-tier/SOA), each server runs identical copies of code (similar to X-axis). The difference here comes in that each server is responsible solely for a strict subset of the data (that is, data partitioning or scaling by splitting into similar things). A given component of the system therefore has responsibility for routing a given request to an appropriate server.

 Sharding is a commonly used routing criteria where an attribute of the request is used to route the request to a particular server (for example, the primary key of a row or identity of a customer).

Just as with X-axis scaling, Z-axis scaling serves to improve application capacity and availability. However, as we learned in this chapter, neither the monolithic or N-tier approach (X- and Y-axis scaling) will resolve the inherent problems of our ever-increasing development and application complexities. To effectively deal with these issues, we need to apply Y-axis scaling (that is, microservices).

This third dimension to scaling (Y-axis) involves functional decomposition, or scaling by splitting into different things. Occurring at the application tier, Y-axis scaling will split a monolithic application into separate sets of services wherein each service implements a set of allied functionalities (for example, customer management, order management, and so on). Later in this chapter, we will look directly into the decomposition of services.

What we can typically see are applications that utilize all the three axes of the scaling cube together. Y-axis scaling decomposes the application into microservices; at runtime, X-axis scaling executes multiple instances of each service behind a load balancer for enhanced output and availability, and some applications may additionally use Z-axis scaling for partition of services.

Disadvantages of microservices

Let's do our full due diligence here by also understanding some of the disadvantages of microservices:

- **Deploying a microservices-based application is much more complex**: In contrast to monolithic applications, a microservice application typically consists of a large number of services. Defacto, we have greater complexity in deploying them.
- **Management and orchestration of microservices is much more complex**: Each service, within a large number of services, will have multiple runtime instances. An exponential increase occurs across many more moving parts that require configuration, deployment, scaling, and monitoring. Any successful microservices deployment, therefore, requires more granular control of deployment methods by developers combined with a high level of automation.
- **Testing a microservices application is much more complex**: Writing test classes for a microservices application does not only require that service to be started, but also its dependency services.

Once understood, we can strategize and design to mitigate these drawbacks and better plan for troubleshooting areas.

Considerations for devising microservices

We have reviewed the transgression from single delivery to multi-tier to containerized microservices, and understand that each has its own functional place for application. Each architecture carries its own degree of validity; appropriate design strategy and application of these architectures is necessary for your deployment successes. Having learned the basic tenets for monolithic, N-tier, and microservices, we are better equipped toward strategically implementing the best-suited architectures on a per use case basis.

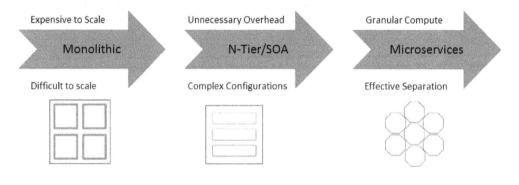

From mono to micro

The microservice architectural pattern is a better choice for complex, evolving applications despite the drawbacks and implementation challenges. To utilize microservices for modern Cloud and web application design and deployment, how best do we leverage the advantages of microservices while mitigating the potential drawbacks?

Whether developing a new application or reinvigorating an old one, these considerations must be taken into account for microservices:

- Building and maintaining highly available distributed systems is complex
- More moving parts means more components to keep track of
- Loosely coupled services means that steps need to be taken to keep data consistent
- Distributed asynchronous processes create network latency and more API traffic
- Testing and monitoring individual services is challenging

Mitigating the disadvantages

This is likely the most simplistic instruction provided in the entire book; however, time and again we witness the obvious either completely ignored, overlooked, or underpursued. Our submission here is that, in spite of the relatively few but known disadvantages, there exist both current and evolving mechanisms to resolve almost all of these issues; expectation is strong that the container market will evolve a plethora of working solutions to the current issues.

Once again, let's just start with the most basic elements here as the foundation of successful microservices applications that require less troubleshooting:

- **Take total ownership**: Without taking full ownership and knowing that ultimate success is directly up to you and your team, your projects and their resulting applications will suffer. Commitment, dedication, and persistence pay handsome results.
- **Develop a complete understanding**: Fully comprehend what the business goals are and what technologies can be best applied to address them, not to mention the *how* and *why* for which you are using them. Always be learning!
- **Pursue exhaustive, coordinated planning**: Plan strategically, plan along with other application stakeholders, plan for failure, and then plan some more; Measure your results and revise the plan, re-evaluating the plan on a continuum. Always be measuring, and always be planning!
- **Utilize the current technology**: It is imperative in today's technology climate to make good use of the most stable and functional tools and applications; so, seek them out.
- **Evolve with the application**: You must be as agile and adaptable as the container technologies you are using; change must be an accepted part of your exhaustive, coordinated planning!

Great! We know that we must not only acknowledge, but actively participate in the most basic elements of our application project process. We also know and understand the advantages and disadvantages of a microservices architectural approach, and that those advantages have the potential to far outweigh any negatives. Outside of the preceding five powerful items, how do we mitigate these drawbacks to use the positives afforded to us with microservices to our benefit?

Managing microservices

At this point, you may be asking yourself "so, where does Docker fit into this conversation?" Our first tongue in cheek answer is that it fits in perfectly, indeed!

Docker is excellent for microservices as it isolates containers to one process or service. This intentional containerization of single services or processes makes it very simple to manage and update these services. Therefore, it's not surprising that the next wave on top of Docker has led to the emergence of frameworks for the sole purpose of managing more complex scenarios, including as follows:

- How to manage single services in a cluster?
- How to manage multiple instances in a service across hosts?
- How to coordinate between multiple services on a deployment and management level?

As expected within a maturing container market, we are seeing additional complementary tools emerge to go along with open source projects, such as Kubernetes, MaestroNG, and Mesos to only name but a few–all arising to address the management, orchestration, and automation needs for containerized applications with Docker. Kubernetes, as an example, is a project built especially for microservices and works extremely well with Docker. The key features of Kubernetes cater directly to the exact traits so imperative within the microservices architecture–easy deployment of new services via Docker, independent scaling of services, end-client transparency to failures, and simple, ad-hoc name-based discovery of service endpoints. Further, Docker's own native projects–Machine, Swarm, Compose, and Orca, while currently still in beta at the time of this writing, look highly promising–will likely soon be added to the Docker core kernel.

Since we will later dedicate examples and discussion to Kubernetes, other third-party applications and an entire chapter to Docker Machine, Swarm, and Compose, let's look at an example here utilizing services we used earlier (NGINX, Node.js) along with Redis and Docker Compose.

Real-world example

NGINX > Node.js > Redis > Docker Compose

```
# Directly create and run the Redis image
docker run -d -name redis -p 6379:6379 redis

## Node Container
# Set the base image to Ubuntu
```

```
FROM ubuntu

# File Author / Maintainer
MAINTAINER John Wooten @CONSULTED <jwooten@cloudconsulted.com>

# Install Node.js and other dependencies
RUN apt-get update && \
        apt-get -y install curl && \
        curl -sL https://deb.nodesource.com/setup | sudo bash - && \
        apt-get -y install python build-essential nodejs

# Install nodemon
RUN npm install -g nodemon

# Provides cached layer for node_modules
ADD package.json /tmp/package.json
RUN cd /tmp && npm install
RUN mkdir -p /src && cp -a /tmp/node_modules /src/

# Define working directory
WORKDIR /src
ADD . /src

# Expose portability
EXPOSE 8080

# Run app using nodemon
CMD ["nodemon", "/src/index.js"]

## Nginx Containers build
# Set nginx base image
FROM nginx

# File Author / Maintainer
MAINTAINER John Wooten @CONSULTED <jwooten@cloudconsulted.com>

# Copy custom configuration file from the current directory
COPY nginx.conf /etc/nginx/nginx.conf

## Docker Compose
nginx:
build: ./nginx
links:
 - node1:node1
 - node2:node2
 - node3:node3
ports:
- "80:80"
```

```
node1:
build: ./node
links:
 - redis
ports:
 - "8080"
node2:
build: ./node
links:
 - redis
ports:
- "8080"
node3:
build: ./node
links:
 - redis
ports:
- "8080"
redis:
image: redis
ports:
 - "6379"
```

We will delve more thoroughly into Docker Compose in `Chapter 10`, *Docker Machine, Compose and Swarm*. Additionally, we will also need to implement a service discovery mechanism (discussed in a later chapter) that enables a service to discover the locations (hosts and ports) of any other services it needs to communicate with.

Automated tests and deployments

We want as much confidence as possible that our applications are working; that starts with automated testing to facilitate our automated deployments. Needless to say, our automated tests are mission-critical. Promotion of working software *up* the pipeline means we automate deployment to each new environment.

The testing of microservices right now is still relatively complex; as we've discussed, test classes for a service will require a launch of that service in addition to any services it depends upon. We at least need to configure stubs for those services. All this can be done, but let's look into mitigating its complexity.

Automated testing

Strategically, we need to map out our design flow to include testing to validate our applications for deployment into production. Here's an example workflow of what we want to accomplish with our automated testing:

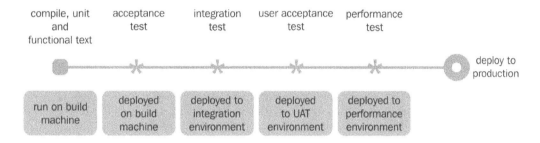

The preceding diagram represents a DevOps pipeline starting with code compilation and moving to integration test, performance test and, finally, the app getting deployed in a production environment.

Designing for failure

In order to succeed, we must accept failures as a very real possibility. In fact, we really ought to be purposefully inserting failures into our application design flow to test how we can successfully deal with them when they occur. This kind of automated testing in production initially requires nerves of steel; however, we can derive automation that is self-healing through repetition and familiarity. Failures are a certainty; therefore, we must plan and test our automation for mitigating the damages of such a certainty.

Successful application design involves built-in fault tolerances; this is particularly true of microservices as a consequence of using services as components. Since services can fail at any time, it's important to be able to detect the failures quickly and, if possible, automatically restore service. Real-time monitoring of our application is of critical emphasis across microservice applications, providing an early warning system of either issues actually occurring or those showing potential for error or problems. This affords an earlier response among development teams to follow up and investigate; because there is such choreography and event collaboration in a microservices architecture, our ability to track emergent behaviors becomes rather vital.

Microservice teams should, therefore, design to include some minimums for monitoring and logging setups for each individual service: dashboards with up/down status, metadata on circuit breaker status, current throughput, and latency and a variety of operational and business relevant metrics.

At the end of our application builds, should our components not compose cleanly, we have accomplished little more than shifting complexity from inside a component to the connections between them. This puts things into areas harder to define and more difficult to control. Ultimately, we should design for the inevitability of failures to be successful.

Dockunit for unit tests

To enhance our unit testing capabilities, we will also install and use **Dockunit** to deliver our unit testing. There are plenty of options available to us for our unit tests. In mixing and matching different tools to accomplish unit testing in the past, I have found that by deploying Dockunit as a *stock and standard* application in my development toolkit, I can meet almost any unit test needs with this utility. So as not to be too repetitive, let's go ahead and set up for automated testing using Dockunit.

Dockunit requirements are Node.js, npm, and Docker.

If not already installed, install npm (we will assume installation of both Docker and Node.js):

```
npm install -g dockunit
```

Now we can use Dockunit to easily test our Node.js applications. This is done simply via a Dockunit.json file; to follow is a sample that tests an application in Node.js 0.10.x and 0.12.0 using mocha:

```
{
  "containers": [
    {
      "prettyName": "Node 0.10.x",
      "image": "google/nodejs:latest",
      "beforeScripts": [
        "npm install -g mocha"
      ],
      "testCommand": "mocha"
    },
    {
      "prettyName": "Node 0.12",
      "image": "tlovett1/nodejs:0.12",
      "beforeScripts": [
        "npm install -g mocha"
```

```
        ],
        "testCommand": "mocha"
      }
    ]
  }
```

The preceding code snippet shows how easily an application can be unit tested inside the docker container.

Automated deployments

One approach to automation is to use an off-the-shelf PaaS (for example, Cloud Foundry or Tutum, and so on). A PaaS provides developers with an easy way to deploy and manage their microservices. It insulates them from concerns such as procuring and configuring IT resources. At the same time, the systems and network professionals who configure the PaaS can ensure compliance with best practices and company policies.

Another way to automate the deployment of microservices is to develop what is essentially your own PaaS. One typical starting point is to use a clustering solution, such as Mesos or Kubernetes, in conjunction with a technology, such as Docker. Later in this book, we will review how software-based application delivery approaches like NGINX, which easily handles caching, access control, API metering, and monitoring at the microservice level can help solve this problem.

Decoupling N-tier applications into multiple images

Decomposing applications improves deployability and scalability and simplifies the adoption of new technologies. To achieve this level of abstraction, the application must be fully decoupled from the infrastructure. Application containers, such as Docker, provide a way to decouple application components from the infrastructure. At this level, each application service must be elastic (that is, it can scale up and down independently of other services) and resilient (that is, it has multiple instances and can survive instance failures). The application should also be designed so that failures in one service do not cascade to other services.

We've done entirely too much talking, and not enough doing. Let's get at what we really need to know–how to build it! We can easily use our `cloudconsulted/wordpress` image here to show an example of our decoupling into separate containers: one for WordPress, PHP, and MySQL. Instead, let's explore other applications to continue to show the range of capabilities and potential for application deployments that we can make with Docker; for this example, a simple LEMP stack

Building an N-tier web application

LEMP stack (NGINX > MySQL > PHP)

For simplification, we will split this LEMP stack across two containers: one for MySQL and the other for NGINX and PHP, each utilizing an Ubuntu base:

```
# LEMP stack decoupled as separate docker container s
FROM ubuntu:14.04
MAINTAINER John Wooten @CONSULTED <jwooten@cloudconsulted.com>
RUN apt-get update
RUN apt-get -y upgrade

# seed database password
COPY mysqlpwdseed /root/mysqlpwdseed
RUN debconf-set-selections /root/mysqlpwdseed
RUN apt-get -y install mysql-server
RUN sed -i -e"s/^bind-address\s*=\s*127.0.0.1/bind-address = 0.0.0.0/" /etc/mysql/my.cnf
RUN /usr/sbin/mysqld & \
    sleep 10s &&\
    echo "GRANT ALL ON *.* TO admin@'%' IDENTIFIED BY 'secret' WITH GRANT OPTION; FLUSH PRIVILEGES" | mysql -u root --password=secret &&\
    echo "create database test" | mysql -u root --password=secret
# persistence:
http://txt.fliglio.com/2013/11/creating-a-mysql-docker-container/
EXPOSE 3306
CMD ["/usr/bin/mysqld_safe"]
```

A second container will install and house NGINX and PHP:

```
# LEMP stack decoupled as separate docker container s
FROM ubuntu:14.04
MAINTAINER John Wooten @CONSULTED <jwooten@cloudconsulted.com>

## install nginx
RUN apt-get update
RUN apt-get -y upgrade
RUN apt-get -y install nginx
RUN echo "daemon off;" >> /etc/nginx/nginx.conf
RUN mv /etc/nginx/sites-available/default /etc/nginx/sites-
available/default.bak
COPY default /etc/nginx/sites-available/default
## install PHP
RUN apt-get -y install php5-fpm php5-mysql
RUN sed -i s/\;cgi\.fix_pathinfo\s*\=\s*1/cgi.fix_pathinfo\=0/
/etc/php5/fpm/php.ini
# prepare php test scripts
RUN echo "<?php phpinfo(); ?>" > /usr/share/nginx/html/info.php
ADD wall.php /usr/share/nginx/html/wall.php
# add volumes for debug and file manipulation
VOLUME ["/var/log/", "/usr/share/nginx/html/"]
EXPOSE 80
CMD service php5-fpm start && nginx
```

Making different tiers of applications work

From our real-world production examples, we have already seen several different ways in which we can make different application tiers work together. Since discussion on making interoperable tiers workable within the application all depend upon the application tiers being deployed, we can continue on *ad-infinitum* as to how to do this; one example leading to another, and so on. Instead, we will delve into this area more thoroughly in `Chapter 06`, *Making Containers Work*.

Summary

Containers are the vehicle for modern microservices architectures; the use of containers provides not some wild and imaginative advantages when coupled with microservices and N-tier architectural styles, but workable production-ready solutions. In many ways, the use of containers to implement a microservices architecture is an evolution not unlike those observed over the past 20 years in web development. Much of this evolution has been driven by the need to make better use of compute resources and the need to maintain increasingly complex web-based applications. For modern application development, Docker is a veritable and forceful weapon.

As we saw, the use of a microservices architecture with Docker containers addresses both these needs. We explored example environments designed seamlessly from development to test, eliminating the need for manual and error-prone resource provisioning and configuration. In doing so, we touched briefly on how a microservice application can be tested, automated, deployed, and managed, but the use of containers in distributed systems goes far beyond microservices. Increasingly, containers are becoming "first class citizens" in all distributed systems and, in the upcoming chapters, we'll discuss how tools such as Docker Compose and Kubernetes are essential for managing container-based computing.

5
Moving Around Containerized Applications

In the last chapter, we covered microservices application architecture deployment with the help of Docker containers. In this chapter, we will explore Docker registry and how it can be used in public and private modes. We will also dive deeply into troubleshooting issues when using public and private Docker registry.

We will look at the following topics:

- Redistributing via Docker registry
- Public Docker registry
- Private Docker registry
- Ensuring integrity of images–signed images
- **Docker Trusted Registry** (**DTR**)
- Docker Universal Control Plane

Redistributing via Docker registry

Docker registry is the server-side application that allows the users to store and distribute Docker images. By default, public Docker registry (Docker Hub) can be used to host multiple Docker images that provides free to use, zero maintenance, and additional features such as automated builds and organization accounts. Let's take a look at public and private Docker registries in detail.

Docker public repository (Docker Hub)

As explained earlier, Docker Hub allows individuals as well as organizations to share the Docker images with its internal teams and customers without the hassle of maintaining a cloud based public repository. It provides centralized resource image discovery and management. It also provides team collaboration and workflow automation for the development pipeline. Some of the additional functions of the Docker Hub, besides Image repository management are as follows:

- **Automated build**: It helps in the creation of new images whenever code is changed in the GitHub or Bitbucket repository
- **WebHooks**: It is a new feature that allows to trigger an action after successful image push to repository
- **User management**: It allows creating workgroups to manage an organization's user access to image repository

An account can be created using the Docker Hub sign-in page in order to host the Docker images; each account will be linked to a unique identification user-based Docker ID. Basic functions, such as Docker image search and *pull* from the Docker Hub, can be performed without creating a Docker Hub account. Images existing in the Docker Hub can be explored using this command:

```
$ docker search centos
```

It will show the existing images in Docker Hub on the basis of the keyword matched.

The Docker ID can also be created using the `docker login` command. The following command will prompt to create a Docker ID that will be public namespace for the user public repository. It will prompt to enter a `Username`, and it will also prompt to enter `Password` and `Email` in order to complete the registration process:

```
$ sudo docker login

Username: username
Password:
Email: email@blank.com
WARNING:login credentials saved in /home/username/.dockercfg.
Account created. Please use the confirmation link we sent to your e-mail to
activate it.
```

In order to log out, the following command can be used:

```
$ docker logout
```

Private Docker registry

Private Docker registry can be deployed inside the local organization; it is open-source under Apache license and is easy to deploy.

Using private Docker registry, you have the following advantages:

- The organization can control and keep a watch on the location where Docker images are stored
- The complete image distribution pipeline will be owned by the organization
- Image storage and distribution will be useful for in-house development workflow and integration with other DevOps components, such as Jenkins

Pushing images to Docker Hub

We can create a customized image that can then be pushed on Docker Hub using tagging. Let's create a simple image with a small terminal-based application. Create a Dockerfile with the following content:

```
FROM debian:wheezy
RUN apt-get update && apt-get install -y cowsay fortune
```

Go to the directory containing the Dockerfile and execute the following command to build an image:

```
$ docker build -t test/cowsay-dockerfile .
Sending build context to Docker daemon 2.048 kB
Sending build context to Docker daemon
Step 0 : FROM debian:wheezy
wheezy: Pulling from debian
048f0abd8cfb: Pull complete
fbe34672ed6a: Pull complete
Digest:
sha256:50d16f4e4ca7ed24aca211446a2ed1b788ab5e3e3302e7fcc11590039c3ab445
Status: Downloaded newer image for debian:wheezy
 ---> fbe34672ed6a
Step 1 : RUN apt-get update && apt-get install -y cowsay fortune
 ---> Running in ece42dc9cffe
```

Alternatively, as shown in the following diagram, we can first create a container and test it out and then create a tagged **Docker Image** that can be easily pushed to **Docker Hub**:

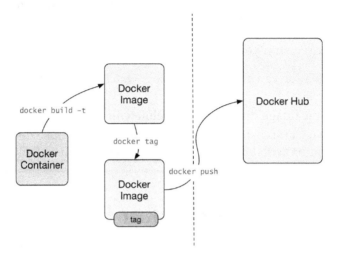

Steps to create a Docker Image from Docker Container and push it to public Docker Hub

We can check whether the image is created using the following command. As you can see, the `test/cowsay-dockerfile` image got created:

```
$ docker images
REPOSITORY                      TAG             IMAGE ID
CREATED               VIRTUAL SIZE
test/cowsay-dockerfile          latest          c1014a025b02        33
seconds ago           126.9 MB
debian                          wheezy          fbe34672ed6a        2
weeks ago             84.92 MB
vkohli/vca-iot-deployment       latest          35c98aa8a51f        8
months ago            501.3 MB
vkohli/vca-cli                  latest          d718bbdc304b        9
months ago            536.6 MB
```

In order to push the image to Docker Hub account, we will have to tag it with the Docker tag/Docker ID using the image ID in the following way:

```
$ docker tag c1014a025b02 username/cowsay-dockerfile
```

As the tagged username will match the Docker Hub ID account, we can easily push the image:

```
$ sudo docker push username/cowsay-dockerfile
The push refers to a repository [username/cowsay-dockerfile] (len: 1)
d94fdd926b02: Image already exists
accbaf2f09a4: Image successfully pushed
aa354fc0b2b2: Image successfully pushed
3a94f42115fb: Image successfully pushed
7771ee293830: Image successfully pushed
fa81ed084842: Image successfully pushed
e04c66a223c4: Image successfully pushed
7e2c5c55ef2c: Image successfully pushed
```

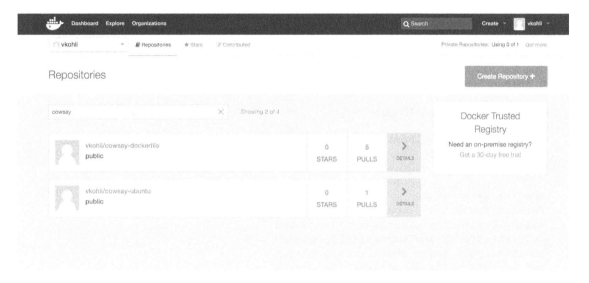

Screenshot of Docker Hub

One of the troubleshooting issues that can be prechecked is that the username tagged on the custom Docker image should meet the username of the Docker Hub account in order to push the image successfully. Custom images pushed to a Docker Hub will be made publicly available. Docker provides one private repository for free, which should be used in order to push the private images. The Docker client version 1.5 and earlier will not be able to push the images to Docker Hub account, but will still be able to pull the images. Only version 1.6 or later are supported. Thus, it is always advised to keep the Docker version up to date.

If the push to Docker Hub fails with a **500 Internal Server Error**, the issue is related to Docker Hub infrastructure and a repush might be helpful. If the issue persists while pushing the Docker image, Docker logs should be referred at `/var/log/docker.log` in in order to debug in detail.

Installing a private local Docker registry

The private Docker registry can be deployed using the image that exists on the Docker Hub. The port mapped to access the private Docker registry will be `5000`:

```
$ docker run -p 5000:5000 registry
```

Now, we will the tag the same image created in the preceding tutorial to `localhost:5000/cowsay-dockerfile` so that the repository name and the image name that match can be easily pushed to private Docker registry:

```
$ docker tag username/cowsay-dockerfile localhost:5000/cowsay-dockerfile
```

Push the image to private Docker registry:

```
$ docker push localhost:5000/cowsay-dockerfile
```

The push refers to a repository (`localhost:5000/cowsay-dockerfile`) (len: 1):

```
Sending image list
Pushing repository localhost:5000/cowsay-dockerfile (1 tags)
e118faab2e16: Image successfully pushed
7e2c5c55ef2c: Image successfully pushed
e04c66a223c4: Image successfully pushed
fa81ed084842: Image successfully pushed
7771ee293830: Image successfully pushed
3a94f42115fb: Image successfully pushed
aa354fc0b2b2: Image successfully pushed
accbaf2f09a4: Image successfully pushed
d94fdd926b02: Image successfully pushed
Pushing tag for rev [d94fdd926b02] on
{http://localhost:5000/v1/repositories/ cowsay-dockerfile/tags/latest}
```

Image ID can be seen by visiting the link in browser or using the `curl` command that comes up after pushing the image.

Moving images in between hosts

Moving an image from one registry to another requires pushing and pulling the image from Internet. If the image is required to be moved from one host to another, then it can be simply achieved with the help of the `docker save` command, without the overhead of uploading and downloading the image. Docker provides two different types of methods in order to save container image to tar ball:

- `docker export`: This saves a container's running or paused state to a tar file
- `docker save`: This saves a non-running container image to a file

Let's compare the `docker export` and `docker save` commands with the help of the following tutorial:

Using export, pull a basic image from Docker Hub:

```
$ docker pull Ubuntu
latest: Pulling from ubuntu
dd25ab30afb3: Pull complete
a83540abf000: Pull complete
630aff59a5d5: Pull complete
cdc870605343: Pull complete
```

Let's create a sample file after running the Docker container from the preceding image:

```
$ docker run -t -i ubuntu /bin/bash
root@3fa633c2e9e6:/# ls
bin  boot  dev  etc  home  lib  lib64  media  mnt  opt  proc  root
run  sbin  srv  sys  tmp  usr  var
root@3fa633c2e9e6:/# touch sample
root@3fa633c2e9e6:/# ls
bin  boot  dev  etc  home  lib  lib64  media  mnt  opt  proc  root
run  sample  sbin  srv  sys  tmp  usr  var
```

In the other shell, we can see the running Docker container and then it can be exported into a tar file using the following command:

```
$  docker ps
CONTAINER ID       IMAGE           COMMAND         CREATED
      STATUS            PORTS           NAMES
3fa633c2e9e6       ubuntu          "/bin/bash"     45 seconds
ago     Up 44 seconds                   prickly_sammet
$ docker export prickly_sammet | gzip > ubuntu.tar.gz
```

The tar file can then be exported to another machine and then imported using the following command:

```
$ gunzip -c ubuntu.tar.gz | docker import - ubuntu-sample
4411d1d3001702b2304d5ebf87f122ef80b463fd6287f3de4e631c50efa01369
```

After we run the container from the Ubuntu-sample image in another machine, we can find the sample file intact:

```
$ docker images
REPOSITORY                      TAG                     IMAGE ID
CREATED
IRTUAL SIZE
ubuntu-sample                       latest                  4411d1d30017      20
seconds
go      108.8 MB

$ docker run -i -t ubuntu-sample /bin/bash
root@7fa063bcc0f4:/# ls
bin  boot  dev  etc  home  lib  lib64  media  mnt  opt  proc  root run
sample
bin  srv  sys  tmp  usr  var
```

Using save, in order to transport the image in spite of the running Docker container as shown in the preceding tutorial, we can use the docker save command that will convert the image into a tar file:

```
$ docker save ubuntu | gzip > ubuntu-bundle.tar.gz
```

The ubuntu-bundle.tar.gz file can now be extracted and used in the other machine using the docker load command:

```
$ gunzip -c ubuntu-bundle.tar.gz | docker load
```

Running the container from the ubuntu-bundle image in the other machine, we will find out that the sample file does not exist as the docker load command will store the image with zero complaints:

```
$ docker run -i -t ubuntu /bin/bash
root@9cdb362f7561:/# ls
bin  boot  dev  etc  home  lib  lib64  media  mnt  opt  proc  root
run  sbin  srv  sys  tmp  usr  var
root@9cdb362f7561:/#
```

Both the preceding examples show the difference between the export and save commands as well as their use in order to transport the images across local hosts without the use of Docker registry.

Ensuring integrity of images – signed images

From Docker version 1.8, the feature included is Docker container trust that integrates **The Update Framework (TUF)** into Docker using Notary, an open source tool which provides trust over any content or data. It allows the verification of the publisher–Docker Engine uses the publisher key in order to verify that–and the image that the user is about to run is exactly what the publisher has created; it has not been tampered with and is up to date. Thus, it is an opt-in feature that allows verification of the publisher of the image. Docker central commands–*push, pull, build, create* and *run*–will operate on the images that either have content signatures or explicit content hashes. The images are signed with private keys by the content publisher before they are pushed to a repository. A trust gets established with publisher when the user interacts with the image for the first time, then all the subsequent interactions require only a valid signature from the same publisher. The model is similar to the first model of SSH that is familiar to us. Docker content trust uses two keys–**offline key** and **tagging key**–which are generated for the first time when the publisher pushes an image. Each repository has its own tagging key. When users run the `docker pull` command for the first time, the trust to repository is established using the offline key:

- **Offline key**: It is the root of trust for your repository; different repositories use the same offline key. This key should be kept offline as it has advantages against certain classes of attacks. Basically, this key is required during creation of a new repository.
- **Tagging key**: It is generated for each new repository that the publisher owns. It can be exported and shared with the person who requires the ability to sign content for the specific repository.

Here's a list of the protection provided by following the trust key structure:

- **Protection against image forgery**: Docker content trust provides protection from man-in-the middle attacks. In case a registry is compromised, the malicious attacker cannot tamper with the content and serve it to users as every run command will fail stating the message unable to verify the content.
- **Protection against reply attacks**: In case of replay attacks, the previous payloads are used by attackers to trick the system. Docker content trust makes use of the timestamp key when publishing the image, thus providing protection against replay attacks and ensuring that a user receives the most up to date content.
- **Protection against key compromise**: The tagging key might get compromised due to its online nature, and it is needed every time new content is pushed to the repository. Docker content trust allows publisher to rotate the compromised key transparently to user and effectively remove it from the system.

Docker content trust is enabled through integration of Notary into Docker Engine. Notary can be downloaded and implemented by anyone who wants to digitally sign and verify arbitrary collection of content. Basically, it is the utility for securely publishing and verifying content over distributed insecure networks. In the following sequence diagram, we can see the flow as to how Notary server is used to verify the metadata files and their integration with Docker client. Trusted collections will be stored in a Notary server and once Docker client has a trusted list of named hashes (tags), it can utilize the Docker remote APIs from client to daemon. Once the pull succeeds, we can trust all the content on manifests and layers in registry pulls.

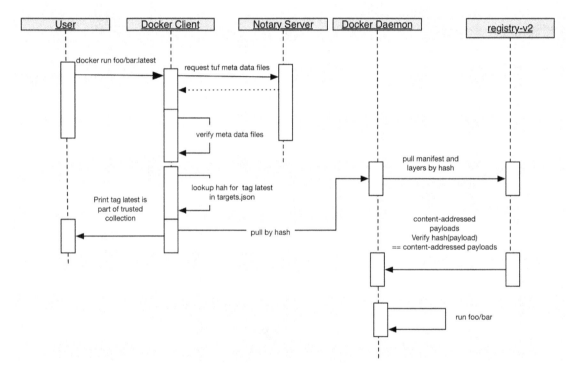

Sequence diagram for Docker trusted run

Internally, Notary uses TUF, a secure general design for software distribution and updates that are often vulnerable to attacks. TUF addresses the widespread problem by providing a comprehensive, flexible-security framework that the developers can integrate with the software update system. Generally, the software update system is an application running on a client system that obtains and installs software.

Let's get started with installing Notary; On Ubuntu 16.04, Notary can be directly installed using this command:

```
$ sudo apt install notary
Reading package lists... Done
Building dependency tree
Reading state information... Done

The following NEW packages will be installed:
  Notary
upgraded, 1 newly installed, 0 to remove and 83 not upgraded.
Need to get 4,894 kB of archives.
After this operation, 22.9 MB of additional disk space will be used.
...
```

Otherwise the project can be downloaded from GitHub and can be manually built and installed; Docker Compose is required to be installed in order to build the project:

```
$ git clone https://github.com/docker/notary.git
Cloning into 'notary'...
remote: Counting objects: 15827, done.
remote: Compressing objects: 100% (15/15), done.

$ docker-compose build
mysql uses an image, skipping
Building signer
Step 1 : FROM golang:1.6.1-alpine


   $ docker-compose up -d
$ mkdir -p ~/.notary && cp cmd/notary/config.json cmd/notary/root-ca.crt
~/.notary
```

After the preceding steps, add `127.0.0.1` Notary server to the `/etc/hosts` or, if a Docker machine is used, add `$(docker-machine ip)` to the Notary server.

Now, we will push the `docker-cowsay` image that we created previously. By default, content trust is disabled; it can be enabled with the help of the `DOCKER_CONTENT_TRUST` environment variable, which will be done later in this tutorial. Currently, the commands that operate with content trust are as shown:

- push
- build
- create
- pull
- run

We will tag the image with the repository name:

```
$ docker images
REPOSITORY                      TAG              IMAGE ID
CREATED              VIRTUAL SIZE
test/cowsay-dockerfile          latest           c1014a025b02       33
seconds ago          126.9 MB
debian                          wheezy           fbe34672ed6a       2
weeks ago            84.92 MB
vkohli/vca-iot-deployment       latest           35c98aa8a51f       8
months ago           501.3 MB
vkohli/vca-cli                  latest           d718bbdc304b       9
months ago           536.6 MB
$ docker tag test/cowsay-dockerfile username/cowsay-dockerfile
$ docker push username/cowsay-dockerfile:latest
The push refers to a repository [docker.io/username/cowsay-dockerfile]
bbb8723d16e2: Pushing 24.08 MB/42.01 MB
```

Now, let's check whether notary has data for this image:

```
$ notary -s https://notary.docker.io -d ~/.docker/trust list
docker.io/vkohli/cowsay-dockerfile:latest
* fatal: no trust data available
```

As we can see here, there is no trust data that lets us enable the `DOCKER_CONTENT_TRUST` flag and then try to push the image:

```
$ docker push vkohli/cowsay-dockerfile:latest
The push refers to a repository [docker.io/vkohli/cowsay-dockerfile]
bbb8723d16e2: Layer already exists
5f70bf18a086: Layer already exists
a25721716984: Layer already exists
latest: digest:
sha256:0fe0af6e0d34217b40aee42bc21766f9841f4dc7a341d2edd5ba0c5d8e45d81c
size: 2609
```

```
Signing and pushing trust metadata
You are about to create a new root signing key passphrase. This passphrase
will be used to protect the most sensitive key in your signing system.
Please
choose a long, complex passphrase and be careful to keep the password and
the
key file itself secure and backed up. It is highly recommended that you use
a
password manager to generate the passphrase and keep it safe. There will be
no
way to recover this key. You can find the key in your config directory.
Enter passphrase for new root key with ID f94af29:
```

As we can see here, for the first time push, it will ask for the passphrase in order to sign the tagged image.

Now we will be getting the trust data from the Notary for the latest image pushed previously:

```
$ notary -s https://notary.docker.io -d ~/.docker/trust list
docker.io/vkohli/cowsay-dockerfile:latest
    NAME
DIGEST                                          SIZE
BYTES)      ROLE
-----------------------------------------------------------------------
-------
--------------------
    latest
0fe0af6e0d34217b40aee42bc21766f9841f4dc7a341d2edd5ba0c5d8e45d81c
1374            targets
```

With the help of the preceding example, we clearly get to know the working of Notary as well as Docker content trust.

Docker Trusted Registry (DTR)

DTR provides enterprise grade Docker image storage on-premises as well as in the virtual private cloud to provide security and meet regulatory compliances. DTR runs on top of Docker **Universal Control Plane** (**UCP**), which can be installed on-premises or on top of the virtual private cloud, with the help of which we can store the Docker images securely behind a firewall.

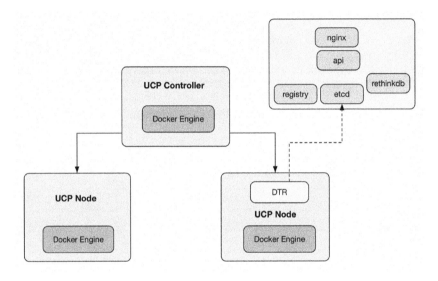

DTR running on UCP node

The two most important features of DTR are as listed:

- **Image management**: It allows the user to store Docker images securely behind firewall and DTR can be easily made as part of the continuous integration and delivery process in order to build, run, and ship applications.

Screenshot of DTR

- **Access control and built-in security**: DTR provides authentication mechanism in order to add users as well as integrates with **Lightweight Directory Access Protocol** (**LDAP**) and Active Directory. It supports **role-based authentication** (**RBAC**) as well, which allows you to assign access control policies for each user.

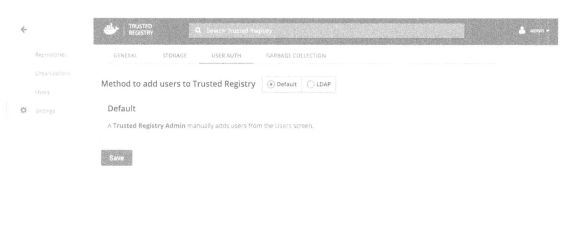

User authentication options in DTR

Docker Universal Control Plane

Docker UCP is the enterprise-grade cluster management solution that allows you to manage the Docker containers from a single platform. It also allows you to manage thousands of nodes, which can be managed and monitored with a graphical UI.

UCP has two important components:

- **Controller**: It manages the cluster and persists the cluster configurations
- **Node**: Multiple nodes can be added to cluster in order to run the containers

UCP can be installed using the sandbox installation on top of Mac OS X or Windows system using **Docker Toolbox**. Installation consists of a UCP controller and one or more hosts that will be added as nodes in the UCP cluster using Docker Toolbox.

A prerequisite for Docker Toolbox is that it is required to be installed for Mac OS X and Windows system using the installer available at the official Docker website.

Docker Toolbox Installation

Let's get started with the deployment of Docker UCP:

1. After the installation, launch the Docker Toolbox terminal:

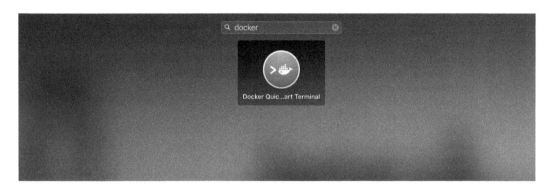

Docker Quickstart Terminal

2. Create a virtual machine named `node1` using the `docker-machine` command and `virtualbox` that will act as UCP controller:

```
$ docker-machine create -d virtualbox --virtualbox-memory
"2000" --virtualbox-disk-size "5000" node1
Running pre-create checks...
Creating machine...
(node1) Copying /Users/vkohli/.docker/machine/cache/
boot2docker.iso to /Users/vkohli/.docker/machine/
machines/node1/boot2docker.iso...
(node1) Creating VirtualBox VM...
(node1) Creating SSH key...
(node1) Starting the VM...
(node1) Check network to re-create if needed...
(node1) Waiting for an IP...
Waiting for machine to be running, this may take a few minutes...
Detecting operating system of created instance...
Waiting for SSH to be available...
Detecting the provisioner...
Provisioning with boot2docker...
Copying certs to the local machine directory...
Copying certs to the remote machine...
Setting Docker configuration on the remote daemon...
Checking connection to Docker...
Docker is up and running!
To see how to connect your Docker Client to the
Docker Engine running on this virtual machine, run:
docker-machine env node1
```

3. Create a `node2` VM as well, which will be configured as a UCP node later:

```
$ docker-machine create -d virtualbox --virtualbox-memory
"2000" node2
Running pre-create checks...

Creating machine...
(node2) Copying /Users/vkohli/.docker/machine/cache/boot2docker.iso
to /Users/vkohli/.docker/machine/machines/node2/
boot2docker.iso...
(node2) Creating VirtualBox VM...
(node2) Creating SSH key...
(node2) Starting the VM...
(node2) Check network to re-create if needed...
(node2) Waiting for an IP...
Waiting for machine to be running, this may take a few minutes...
Detecting operating system of created instance...
Waiting for SSH to be available...
Detecting the provisioner...
Provisioning with boot2docker...
Copying certs to the local machine directory...
Copying certs to the remote machine...
Setting Docker configuration on the remote daemon...
Checking connection to Docker...
Docker is up and running!
To see how to connect your Docker Client to the
Docker Engine running on this virtual machine,
run: docker-machine env node2
```

4. Configure `node1` as a UCP controller, which will be responsible for serving the UCP application and running the processes to manage Docker objects' installation. Before that, set the environment to configure `node1` as a UCP controller:

```
$ docker-machine env node1
export DOCKER_TLS_VERIFY="1"
export DOCKER_HOST="tcp://192.168.99.100:2376"
export
DOCKER_CERT_PATH="/Users/vkohli/.docker/machine/machines/node1"
export DOCKER_MACHINE_NAME="node1"
# Run this command to configure your shell:
# eval $(docker-machine env node1)

$ eval $(docker-machine env node1)

$ docker-machine ls
```

```
NAME     ACTIVE    DRIVER       STATE    URL          SWARM
DOCKER   ERRORS
node1    *         virtualbox   Running
tcp://192.168.99.100:2376
1.11.1
node2    -         virtualbox   Running           v1.11.1
tcp://192.168.99.101:2376
```

5. While setting the `node1` as a UCP controller, it will ask for the password for the UCP admin account and additional aliases will be asked for, which can be added or skipped with the enter command:

```
$ docker run --rm -it -v /var/run/docker.sock:/var/run
/docker.sock --name ucp docker/ucp install -i --swarm-port
3376 --host-address $(docker-machine ip node1)

Unable to find image 'docker/ucp:latest' locally
latest: Pulling from docker/ucp
...
Please choose your initial UCP admin password:
Confirm your initial password:
INFO[0023] Pulling required images... (this may take a while)
WARN[0646] None of the hostnames we'll be using in the UCP
certificates [node1 127.0.0.1 172.17.0.1 192.168.99.100]
contain a domain component.  Your generated certs may fail
TLS validation unless you only use one of these shortnames
or IPs to connect.  You can use the --san flag to add more aliases

You may enter additional aliases (SANs) now or press enter to
proceed with the above list.
Additional aliases: INFO[0646] Installing UCP with host address
192.168.99.100 - If this is incorrect, please specify an
alternative address with the '--host-address' flag
INFO[0000] Checking that required ports are available and
accessible

INFO[0002] Generating UCP Cluster Root CA
INFO[0039] Generating UCP Client Root CA
INFO[0043] Deploying UCP Containers
INFO[0052] New configuration established.  Signalling the daemon
to load it...
INFO[0053] Successfully delivered signal to daemon
INFO[0053] UCP instance ID:
KLIE:IHVL:PIDW:ZMVJ:Z4AC:JWEX:RZL5:U56Y:GRMM:FAOI:PPV7:5TZZ
INFO[0053] UCP Server SSL: SHA-256
Fingerprint=17:39:13:4A:B0:D9:E8:CC:31:AD:65:5D:
```

```
52:1F:ED:72:F0:81:51:CF:07:74:85:F3:4A:66:F1:C0:A1:CC:7E:C6
INFO[0053] Login as "admin"/(your admin password) to UCP at

https://192.168.99.100:443
```

6. The UCP console can be accessed using the URL provided at the end of installation; log in with `admin` as the username and the password that you set previously while installing.

Docker UCP license page

7. After logging in, the trail license can be added or skipped. The trail license can be downloaded by following the link on the UCP dashboard on the Docker website. The UCP console with multiple options such as listing application, container, and nodes:

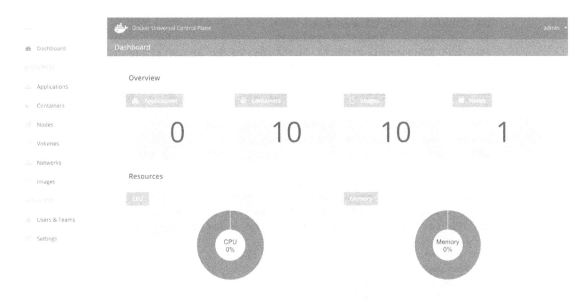

Docker UCP management dashboard

8. Join UCP `node2` to the controller first by setting the environment:

```
$ docker-machine env node2
export DOCKER_TLS_VERIFY="1"
export DOCKER_HOST="tcp://192.168.99.102:2376"
export
DOCKER_CERT_PATH="/Users/vkohli/.docker/machine/machines/node2"
export DOCKER_MACHINE_NAME="node2"
# Run this command to configure your shell:
# eval $(docker-machine env node2)
$ eval $(docker-machine env node2)
```

9. Add the node to UCP controller using the following command. UCP controller URL, username, and password will be asked for, as illustrated:

```
$ docker run --rm -it -v /var/run/docker.sock:/var/run/docker.sock
  --name ucp docker/ucp join -i --host-address
$(docker-machine ip node2)

Unable to find image 'docker/ucp:latest' locally
latest: Pulling from docker/ucp
...

Please enter the URL to your UCP server: https://192.168.99.101:443
UCP server https://192.168.99.101:443
```

```
CA Subject: UCP Client Root CA
Serial Number: 4c826182c994a42f
SHA-256 Fingerprint=F3:15:5C:DF:D9:78:61:5B:DF:5F:39:1C:D6:
CF:93:E4:3E:78:58:AC:43:B9:CE:53:43:76:50:
00:F8:D7:22:37
Do you want to trust this server and proceed with the join?
(y/n): y
Please enter your UCP Admin username: admin
Please enter your UCP Admin password:
INFO[0028] Pulling required images... (this may take a while)
WARN[0176] None of the hostnames we'll be using in the UCP
certificates [node2 127.0.0.1 172.17.0.1 192.168.99.102]
contain a domain component.  Your generated certs may fail
TLS validation unless you only use one of these shortnames
or IPs to connect.  You can use the --san flag to add more aliases

You may enter additional aliases (SANs) now or press enter
to proceed with the above list.
Additional aliases:
INFO[0000] This engine will join UCP and advertise itself
with host address 192.168.99.102 - If this is incorrect,
please specify an alternative address with the '--host-address'
flag
INFO[0000] Verifying your system is compatible with UCP
INFO[0007] Starting local swarm containers
INFO[0007] New configuration established.  Signalling the
daemon to load it...
INFO[0008] Successfully delivered signal to daemon
```

10. The installation of UCP is complete; now DTR can be installed on node2 by pulling the official DTR image from Docker Hub. UCP URL, username, password, and certificate will also be required in order to complete the DTR installation:

```
$ curl -k https://192.168.99.101:443/ca > ucp-ca.pem

$ docker run -it --rm docker/dtr install --ucp-url https://
192.168.99.101:443/ --ucp-node node2 --dtr-load-balancer
192.168.99.102 --ucp-username admin --ucp-password 123456
--ucp-ca "$(cat ucp-ca.pem)"

INFO[0000] Beginning Docker Trusted Registry installation
INFO[0000] Connecting to network: node2/dtr-br
INFO[0000] Waiting for phase2 container to be known to the
Docker daemon
INFO[0000] Connecting to network: dtr-ol
...
```

```
INFO[0011] Installation is complete
INFO[0011] Replica ID is set to: 7a9b6eb67065
INFO[0011] You can use flag '--existing-replica-id 7a9b6eb67065'
when joining other replicas to your Docker Trusted Registry Cluster
```

11. After the successful installation, DTR can be listed as an application in the UCP UI:

Docker UCP listing all the applications

12. The DTR UI can be accessed using the `http://node2` URL. The new repository can be created by clicking on the **New repository** button:

Creating a new private registry in DTR

13. The images can be pushed and pulled from the secured DTR created previously and the repository can be made private as well in order to keep the internal company-wide containers secured.

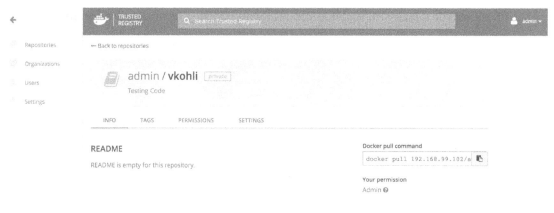

Creating a new private registry in DTR

14. DTR can be configured using the **Settings** option from the menu that allows to set the domain name, TLS certificate, and storage backend for Docker images.

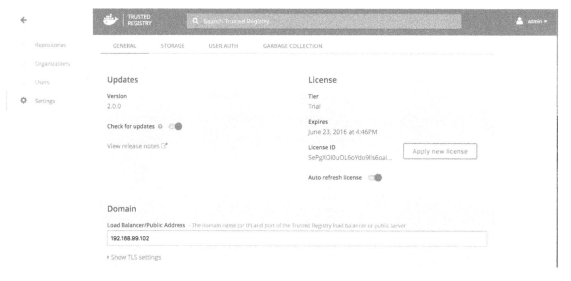

Settings option in DTR

Summary

In this chapter, we dived deeply into Docker registry. We started with the basic concepts of a Docker public repository using Docker Hub and the use-case of sharing containers with a larger audience. Docker also provides the option to deploy a private Docker registry that we looked into and that can be used to push, pull, and share the Docker containers internally in the organization. Then, we looked into tagging and ensuring the integrity of Docker containers by signing them with the help of a Notary server, which can be integrated with Docker Engine. A more robust solution is provided with the help of DTR, which provides enterprise grade Docker image storage on-premises as well as in the virtual private cloud to provide security and meet regulatory compliances. It runs on top of the Docker UCP, as shown in the preceding detailed installation steps. I hope this chapter has helped you troubleshoot and learn the latest trends in Docker registry. In the next chapter, we will look into making containers work with the help of privileged containers and their resource sharing.

6
Making Containers Work

In this chapter, we will explore various options of creating Docker containers with added modes, such as privileged mode and super privileged mode containers. We will also be exploring various troubleshooting issues for these modes.

We will take a deep dive into various deployment management tools, such as **Chef**, **Puppet**, and **Ansible**, which provide integration with Docker in order to ease the pain of deploying thousands of containers for a production environment.

In this chapter, we will cover the following topics:

- Privileged containers and super privileged containers
- Troubleshooting issues of working with different sets of options available for creating containers
- Making Docker containers work with Puppet, Ansible, and Chef
- Using Puppet to create Docker containers and deploy applications
- Managing Docker containers with Ansible
- Building Docker and Ansible together
- Chef for Docker

Automating the Docker container's deployment with the help of the preceding management tools has the following advantages:

- **Flexibility**: They provide you with the flexibility to reproduce the Docker-based application, as well as the environment required for the Docker application on the cloud instance or bare metal of your choice. This helps in managing and testing, as well as providing dev environment spin up as and when required.
- **Auditability**: These tools also provide auditability, as they provide isolation and help track any potential vulnerabilities and who deployed what type of container in which environment.
- **Ubiquity**: They help you manage the full environment around containers, that is, manage container as well as non-container environments such as storage, database, and networking models around the container application.

Privileged containers

By default, containers run in unprivileged mode, that is, we cannot run Docker daemon inside a Docker container. However, the privileged Docker container is given access to all the devices. Docker privileged mode allows access to all the devices on the host and sets system configuration in **App Armor** and **SELinux** to allow containers the same access as the process running on the host:

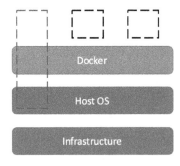

Privileged container highlighted in red

The privileged containers can be started with the following command:

```
$ docker run -it --privileged ubuntu /bin/bash
root@9ab706a6a95c:/# cd /dev/
root@9ab706a6a95c:/dev# ls
agpgart         hdb6              psaux   sg1        tty32   tty7
atibm           hdb7              ptmx    shm        tty33   tty8
audio           hdb8              pts     snapshot   tty34   tty9
beep            hdb9              ram0    sr0        tty35   ttyS0
```

As we can see, after starting the container in privileged mode, we can list all the devices connected to the host machine.

Troubleshooting tips

Docker allows you to use the non-default profile by supporting the addition, as well as the removal of, capabilities. It is better to remove the capabilities that are not specifically required for the container process as this will make it secure.

If you are facing security threats on your host system running containers, it is usually advised to check if any of the containers are running with privileged mode, which might be affecting the security of the host system by running a security-threat application.

As seen in the following example, when we run the container without privileged mode, we are unable to change the kernel parameters, but when we run the container in privileged mode using the `--privileged` flag it is able to change the kernel parameters easily, which can cause a security vulnerability on the host system:

```
$ docker run -it centos /bin/bash
[root@7e1b1fa4fb89 /]#  sysctl -w net.ipv4.ip_forward=0
sysctl: setting key "net.ipv4.ip_forward": Read-only file system
$ docker run --privileged -it centos /bin/bash
[root@930aaa93b4e4 /]#  sysctl -a | wc -l
sysctl: reading key "net.ipv6.conf.all.stable_secret"
sysctl: reading key "net.ipv6.conf.default.stable_secret"
sysctl: reading key "net.ipv6.conf.eth0.stable_secret"
sysctl: reading key "net.ipv6.conf.lo.stable_secret"
638
[root@930aaa93b4e4 /]# sysctl -w net.ipv4.ip_forward=0
net.ipv4.ip_forward = 0
```

So, while auditing, you should ensure that all the containers running on the host system do not have privileged mode set to `true` unless required for some specific application running in the Docker container:

```
$ docker ps -q | xargs docker inspect --format '{{ .Id }}:
Privileged={{
.HostConfig.Privileged }}'
930aaa93b4e44c0f647b53b3e934ce162fbd9ef1fd4ec82b826f55357f6fdf3a:
Privileged=true
```

Super-privileged container

This concept is introduced in one of the Project Atomic blogs, by Redhat. It provides the capability to use a special/privileged container as an agent to control the underlying host. If we ship only the application code, we risk turning the container into a black box. There are many benefits to the host of packaging up an agent as a Docker container with the right access. We can bind in devices via `-v /dev:/dev`, which will help to mount devices inside the container without needing super-privileged access.

Using `nsenter` trick, allows you to run commands in another namespace, that is, if Docker has its own private mount namespace, with `nsenter` and the right mode we can reach out to the host and mount things in its namespace.

We can run in privileged mode to mount the whole host system on some path (`/media/host`):

```
$ docker run -it -v /:/media/host --privileged fedora
nsenter --mount=/media/host/proc/1/ns/mnt --mount /dev/xvdf /home/mic
```

We can then use `nsenter` inside the container; `--mount` tells `nsenter` to look into `/media/host` and then select the mount namespace for proc number 1. Then, run the regular mount command linking the device to the mount point. As seen previously, this functionality allows us to mount host sockets and devices such as a file, and thus all can be bind mounted into a container for use:

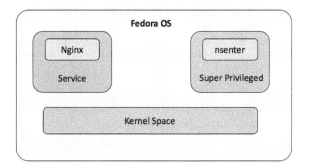

nsenter running as a super-privileged container monitoring the host

Basically, super-privileged containers thus not only provide security separation, resource, and process isolation, but also a mechanism for shipping containers. Allowing software to be shipped as a container image also allows us to manage the host operating system and manage other container processes as explained previously.

Let us consider an example where, currently, we are loading the required kernel modules as RPM packages needed by the application that are not included in the host OS, and running them when the application starts. This module can be shipped with the help of super-privileged containers, and the benefit will be that this custom kernel module can work very well with the current kernel in comparison to shipping kernel modules as part of a privileged container. In this approach, it is not required to run the application as a privileged container; they can run separately and kernel modules can be loaded as part of a different image as shown here:

```
$ sudo docker run --rm --privileged foobar /sbin/modprobe PATHTO/foobar-kmod
$ sudo docker run -d foobar
```

Troubleshooting – Docker containers at scale

Working in a production environment means continuous deployments. When the infrastructure is decentralized and cloud-based, we are frequently managing the deployment of identical services across identical systems. Automating the entire process of configuration and management of this system will be a boon. Deployment management tools are designed for this purpose. They provide recipes, playbooks, and templates to simplify orchestration and automation, to provide a standard and consistent deployment. In the following sections, we will be exploring three common configuration-automation tools: Chef, Puppet, and Ansible, and the ease they provide for deploying Docker containers at scale.

Puppet

Puppet is an automated engine that performs automated administrative tasks such as updating configurations, adding users, and installing packages based on user specifications. Puppet is a well known open source configuration management tool, which runs on various systems, such as Microsoft Windows, Unix, and Linux. The user describes the configuration using either Puppet's declarative language or a domain-specific language (Ruby). Puppet is model-driven and requires limited programming knowledge to use. Puppet provides a module for managing Docker containers. Puppet and Docker integration can help to achieve complex use cases with ease. Puppet manages files, packages, and services, while Docker encapsulates binaries and configuration inside a container, for deployment as an application.

One of the potential use cases of Puppet is that it can be used to provision the Docker containers required for a Jenkins build, and this can be done at scale as per the need of developers, that is, when the build gets triggered. After the build process is complete, binaries can be delivered to the respective owners and containers can be destroyed after each build. Puppet plays a very important role in this use case as the code has to be written once using the Puppet template, and it can be triggered as and when required:

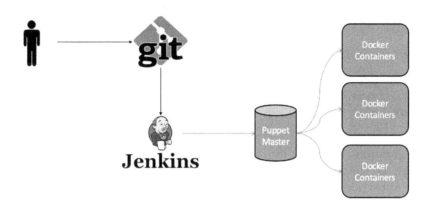

Integration of Puppet and Jenkins to deploy build docker containers

The Puppet module for managing Docker can be installed as per the `garethr-docker` GitHub project. The module just requires a single class to be included:

```
include 'docker'
```

It sets up a Docker hosted repository and installs Docker packages and any required kernel extensions. The Docker daemon will bind to `unix socket /var/run/docker.sock`; this configuration can be changed as per the requirement:

```
class { 'docker':
  tcp_bind        => ['tcp://127.0.0.1:4245','tcp://10.0.0.1:4244'],
  socket_bind     => 'unix:///var/run/docker.sock',
  ip_forward      => true,
  iptables        => true,
  ip_masq         => true,
  bridge          => br0,
  fixed_cidr      => '10.21.1.0/24',
  default_gateway => '10.21.0.1',
}
```

As shown in the preceding code, the default configuration the Docker can be changed as per the configurations provided by this module.

Images

The Docker image can be pulled with the help of the configurations syntax elaborated here.

The alternative to the `ubuntu:trusty docker` command will be as follows:

```
$ docker pull -t="trusty" ubuntu
docker::image { 'ubuntu':
  image_tag => 'trusty'
}
```

Even the configuration allows the link to Dockerfile in order to build the image. A rebuild of the image can also be triggered by subscribing to external events such as changes in the Dockerfile. We subscribe to changes in the folder `vkohli/Dockerfile`, as follows:

```
docker::image { 'ubuntu':
  docker_file => '/vkohli/Dockerfile'
  subscribe => File['/vkohli/Dockerfile'],
}
file { '/vkohli/Dockerfile':
  ensure => file,
  source => 'puppet:///modules/someModule/Dockerfile',
}
```

Containers

After the image has been created, containers can be launched with a number of optional parameters. We get a similar functionality with the basic `docker run` command:

```
docker::run { 'sampleapplication':
  image          => 'base',
  command        => '/bin/sh -c "while true; do echo hello world;
sleep 1;
                    done"',
  ports          => ['4445', '4555'],
  expose         => ['4665', '4777'],
  links          => ['mysql:db'],
  net            => 'my-user-def',
  volumes        => ['/var/lib/couchdb', '/var/log'],
  volumes_from   => '6446ea52fbc9',
  memory_limit   => '20m', # (format: '<number><unit>', where unit =
b, k, m
                    or g)
  cpuset         => ['0', '4'],
  username       => 'sample',
  hostname       => 'sample.com',
```

```
    dns              => ['8.8.8.8', '8.8.4.4'],
    restart_service => true,
    privileged       => false,
    pull_on_start    => false,
    before_stop      => 'echo "The sample application completed"',
    after            => [ 'container_b', 'mysql' ],
    depends          => [ 'container_a', 'postgres' ],
    extra_parameters => [ '--restart=always' ],
}
```

As shown here, we are also able to pass some more parameters, such as the following:

- `pull_on_start`: Before the image is started it will be freshly pulled each time
- `before_stop`: The command mentioned will get executed before stopping the container
- `extra_parameters`: Additional array parameters required to pass to the `docker run` command, such as `--restart=always`
- `after`: This option allows expressing containers that are required to be started first

Other parameters which can be set are `ports`, `expose`, `env_files`, and `volumes`. A single value or an array of values can be passed.

Networking

The latest Docker versions have official support for networks: the module now exposes a type, Docker network, which can be used to manage them:

```
docker_network { 'sample-net':
    ensure   => present,
    driver   => 'overlay',
    subnet   => '192.168.1.0/24',
    gateway  => '192.168.1.1',
    ip_range => '192.168.1.4/32',
}
```

As the preceding code shows, a new overlay network, `sample-net`, can be created, and the Docker daemon can be configured to use it.

Docker compose

Compose is a tool for running multiple Docker container applications. Using the compose file, we can configure an application's services and start them as well. The `docker_compose` module type is provided, which allows Puppet to run the compose application with ease.

A compose file can be added as well, such as scaling rules of running four containers, as shown in the following code snippet. We can also provide additional parameters required for networking and other configurations:

```
docker_compose { '/vkohli/docker-compose.yml':
  ensure  => present,
  scale   => {
    'compose_test' => 4,
  },
  options => '--x-networking'
}
```

1. If the Puppet program is not installed on your machine it can be done in the following way:

   ```
   $ puppet module install garethr-docker
   The program 'puppet' is currently not installed. On Ubuntu 14.04
   the
   puppet program
   can be installed as shown below;
   $ apt-get install puppet-common
   Reading package lists... Done
   Building dependency tree
   Reading state information... Done
   ...
   The following extra packages will be installed:
   Unpacking puppet-common (3.4.3-1ubuntu1.1) ...
   Selecting previously unselected package ruby-rgen.
   Preparing to unpack .../ruby-rgen_0.6.6-1_all.deb ...
   ...
   ```

2. After the Puppet module installation, the `garethr-docker` module can be installed as shown:

   ```
   $ puppet module install garethr-docker
   Notice: Preparing to install into /etc/puppet/modules ...
   Notice: Downloading from https://forge.puppetlabs.com ...
   Notice: Installing -- do not interrupt ...
   /etc/puppet/modules
   ```

```
|__ garethr-docker (v5.3.0)
  |__ puppetlabs-apt (v2.2.2)
  |__ puppetlabs-stdlib (v4.12.0)
  |__ stahnma-epel (v1.2.2)
```

3. We will be creating one sample hello world app, which will be deployed using Puppet:

```
$ nano sample.pp
include 'docker'
docker::image { 'ubuntu':
  image_tag => 'precise'
}
docker::run { 'helloworld':
  image => 'ubuntu',
  command => '/bin/sh -c "while true; do echo hello world; sleep 1;
          done"',
}
```

4. After creating the file, we apply (run) it:

```
$ puppet apply sample.pp
Warning: Config file /etc/puppet/hiera.yaml not found, using Hiera
defaults
Warning: Scope(Apt::Source[docker]): $include_src is deprecated and
will be removed in the next major release, please use $include => {
'src' => false } instead
. . .
Notice: /Stage[main]/Main/Docker::Run[helloworld]/Service[docker-
helloworld]/ensure:
ensure changed 'stopped' to 'running'
Notice: Finished catalog run in 0.80 seconds
Post installation it can be listed as running container:
$ docker ps
CONTAINER ID        IMAGE             COMMAND
CREATED             STATUS            PORTS             NAMES
bd73536c7f64        ubuntu:trusty     "/bin/sh -c 'while tr"    5
seconds ago         Up 5 seconds      helloworld
```

5. We can attach it to the container and see the output:

```
$ docker attach bd7
hello world
hello world
hello world
hello world
```

As we have shown earlier, containers can be deployed across multiple hosts and the entire cluster can be created with help of single Puppet configuration file.

Troubleshooting tips

If you are not able to list the Docker image even after the Puppet `apply` command has run successfully, check the syntax and whether the correct image name is put up in the sample file.

Ansible

Ansible is a workflow orchestration tool that provides configuration management, provisioning, and application deployment with the help of one easy-to-use platform. Some of the powerful features of Ansible are as follows:

- **Provisioning**: The apps are developed and deployed in different environments. It can be on bare metal servers, VMs, or Docker containers, locally or on the cloud. Ansible can help to streamline the provisioning steps with the help of Ansible tower and playbooks.
- **Configuration Management**: Keeping a common configuration file is one of the primary use cases of Ansible, and helps manage and deploy in the required environment.
- **Application Deployment**: Ansible helps to manage the complete lifecycle of an application, from deployment to production.
- **Continuous Delivery**: Managing a continuous delivery pipeline requires resources from various teams. It cannot be achieved with the help of simple platform, hence, Ansible playbooks play a vital role in deploying and managing the applications throughout their lifecycle.
- **Security and Compliance**: Security can be an integral part from the deployment stage, by integrating various security policies as part of the automated process, rather than as an afterthought process or merging it later.
- **Orchestration**: As explained previously, Ansible can define the way to manage multiple configurations, interact with them, and manage the individual pieces of the deployment script.

Automating Docker with Ansible

Ansible also provides a way to automate Docker containers; it enables us to channelise and operationalise the Docker container build and automate a process that is mostly handled manually as of now. Ansible offers the following module for orchestrating Docker containers:

- **Docker_service**: The existing Docker compose files can be used to orchestrate containers on a single Docker daemon or swarm with the help of the Docker service part of Ansible. The Docker compose file has the same syntax as the Ansible playbook, as both of them are **Yaml** files and the syntax is almost the same. Ansible is also written in Python, and the Docker module uses the exact docker-py API client that docker compose uses internally.

Here's a simple Docker compose file:

```
wordpress:
image: wordpress
links:
    - db:mysql
ports:
    - 8080:80
db:
image: mariadb
environment:
        MYSQL_ROOT_PASSWORD: sample
```

The Ansible playbook for the preceding Docker compose file looks similar:

```
# tasks file for ansible-dockerized-wordpress
- name: "Launching DB container"
 docker:
   name: db
   image: mariadb
   env:
     MYSQL_ROOT_PASSWORD: esample
- name: "Launching wordpress container"
 docker:
   name: wordpress
   image: wordpress
   links:
   - db:mysql
   ports:
   - 8081:80
```

- **docker_container**: This manages the lifecycle of the Docker container by providing the ability to start, stop, create, and destroy a Docker container.
- **docker_image**: This provides help to manage images of the Docker container with commands such as build, push, tag, and remove a Docker image.
- **docker_login**: This authenticates with the Docker hub or any Docker registry and provides pushing as well as pulling Docker images from the registry.

Ansible Container

Ansible Container is a tool used to orchestrate and build Docker images using Ansible playbooks only. Ansible Container can be installed in the following way by creating `virtualenv` using pip installation:

```
$ virtualenv ansible-container
New python executable in /Users/vkohli/ansible-container/bin/python
Installing setuptools, pip, wheel...done.
vkohli-m01:~ vkohli$ source ansible-container/bin/activate
(ansible-container) vkohli-m01:~ vkohli$ pip install ansible-container
Collecting ansible-container
  Using cached ansible-container-0.1.0.tar.gz
Collecting docker-compose==1.7.0 (from ansible-container)
  Downloading docker-compose-1.7.0.tar.gz (141kB)
    100% |==============================| 143kB 1.1MB/s
Collecting docker-py==1.8.0 (from ansible-container)
  ...
  Downloading docker_py-1.8.0-py2.py3-none-any.whl (41kB)
  Collecting cached-property<2,>=1.2.0 (from docker-
compose==1.7.0->ansible-
  container)
```

Troubleshooting tips

If you have issues installing Ansible Container as shown above, the installation can be done by downloading the source code from GitHub:

```
$ git clone https://github.com/ansible/ansible-container.git
Cloning into 'ansible-container'...
remote: Counting objects: 2032, done.
remote: Total 2032 (delta 0), reused 0 (delta 0), pack-reused 2032
Receiving objects: 100% (2032/2032), 725.29 KiB | 124.00 KiB/s, done.
Resolving deltas: 100% (1277/1277), done.
Checking connectivity... done.
$ cd ansible-container/
```

```
$ ls
AUTHORS        container        docs      EXAMPLES.md  LICENSE
README.md           setup.py  update-authors.py
codecov.yml  CONTRIBUTORS.md  example  INSTALL.md   MANIFEST.in
requirements.txt  test
$ sudo python setup.py install
running install
running bdist_egg
running egg_info
creating ansible_container.egg-info
writing requirements to ansible_container.egg-info/requires.txt
```

The Ansible Container has the following commands to get started:

- **ansible_container init**: This command creates a directory for Ansible files to get started:

  ```
  $ ansible-container init
  Ansible Container initialized.
  $ cd ansible
  $ ls
  container.yml    main.yml    requirements.tx
  ```

- **ansible-container build**: This creates images from the Ansible playbooks in the Ansible directory

- **ansible-container run**: This launches the containers defined in the `container.yml` file

- **ansible-container push**: This pushes the project's image to the private or public repository, as per the user's choice

- **ansible-container shipit**: This will export the necessary playbooks and roles to deploy containers to a supported cloud provider

As shown in the example at GitHub, the Django service can be defined in the `container.yml` file in the following way:

```
version: "1"
services:
  django:
    image: centos:7
    expose:
      - "8080"
    working_dir: '/django'
```

Chef

Chef has some important components, such as cookbook and recipes. A cookbook defines a scenario and contains everything; the first of them is recipes which is a fundamental configuration element within an organisation and it is written in Ruby language. It is mostly collection of resource-defined using patterns. Cookbooks also contain attribute values, file distribution, and templates. Chef allows the Docker container to be managed in a versionable, testable, and repeatable way. It provides you with the power to build an efficient workflow for container-based development and to manage the release pipeline. Chef delivery allows you to automate and use the scalable workflow to test, develop, and release the Docker container.

The Docker cookbook is available on GitHub (https://github.com/chef-cookbooks/docker) and provides custom resources to be used in the recipes. It provides various options, such as the following:

- docker_service: These are the composite resources used for docker_installation and docker_service manager
- docker_image: This deals with pulling Docker images from a repository
- docker_container: This handles all the Docker container operations
- docker_registry: This handles all the Docker registry operations
- docker_volume: This manages all the volume related operations for Docker containers

The following is a sample Chef Docker recipe, which can be used for reference to deploy the containers using Chef recipes:

```
# Pull latest nginx image
docker_image 'nginx' do
  tag 'latest'
  action :pull
  notifies :redeploy, 'docker_container[sample_nginx]'
end
# Run container by exposing the ports
docker_container 'sample_nginx' do
  repo 'nginx'
  tag 'latest'
  port '80:80'
  host_name 'www'
  domain_name 'computers.biz'
  env 'FOO=bar'
  volumes [ '/some/local/files/:/etc/nginx/conf.d' ]
end
```

Summary

In this chapter, we initially did a deep dive into privileged containers, which can get access to all the host devices as well as super-privileged containers, it shows the capability of the containers to manage to run a background service which can be used to run services in Docker containers to manage the underlying host. Then, we looked into Puppet, a key orchestration tool, and how it handles container management with help of the `garethr-docker` GitHub project. We also looked into Ansible and Chef, which provide similar capabilities to Puppet to manage Docker containers at scale. In the next chapter, we will be exploring the Docker networking stack.

7
Managing the Networking Stack of a Docker Container

In this chapter, we will cover the following topics:

- docker0 bridge
- Troubleshooting Docker bridge configuration
- Configuring DNS
- Troubleshooting communication between containers and the external network
- ibnetwork and the Container Network Model
- Docker networking tools based on overlay and underlay networks
- Comparison of Docker networking tools
- Configuring **OpenvSwitch** (**OVS**) to work with Docker

Docker networking

Each Docker container has its own network stack, and this is due to the Linux kernel `net` namespace, where a new `net` namespace for each container is instantiated and cannot be seen from outside the container or other containers.

Docker networking is powered by the following network components and services:

- **Linux bridges**: L2/MAC learning switch built into the kernel to use for forwarding
- **Open vSwitch**: Advanced bridge that is programmable and supports tunneling
- **Network Address Translators (NAT)**: These are immediate entities that translate IP address + Ports (SNAT, DNAT)

- **IPtables**: Policy engine in the kernel that is used for managing packet forwarding, firewall, and NAT features
- **Apparmor/SElinux**: Firewall policies for each application can be defined

Various networking components can be used to work with Docker, providing new ways to access and use Docker-based services. As a result, we see a lot of libraries that follow different approaches to networking. Some prominent ones are Docker Compose, Weave, Kubernetes, Pipework, and libnetwork. The following diagram depicts root ideas of Docker networking:

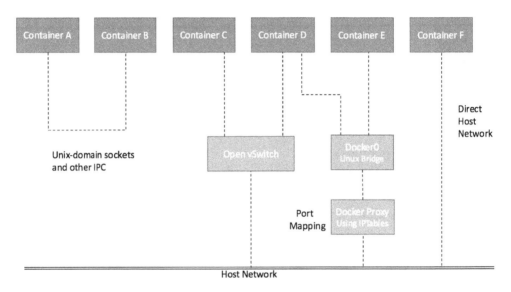

Docker networking modes

docker0 bridge

docker0 bridge is the heart of default networking. When the Docker service is started, a Linux bridge is created on the host machine. The interfaces on the containers talk to the bridge and the bridge proxies to the external world. Multiple containers on the same host can talk to each other through the Linux bridge.

docker0 can be configured via the `--net` flag, and has four modes in general:

- `--net default`: In this mode, the default bridge is used as the bridge for containers to connect to each other
- `--net=none`: With this flag, the container created is truly isolated and cannot connect to the network
- `--net=container:$container2`: With this flag, the container created shares its network namespace with the container named `$container2`
- `--net=host`: In this mode, the container created shares its network namespace with the host

Troubleshooting Docker bridge configuration

In this section, we will look at how the container ports are mapped to host ports and how we can troubleshoot the issue of connecting containers to the external world. This mapping can be done either implicitly by the Docker Engine or can be specified.

If we create two containers–**Container 1** and **Container 2**–both of them are assigned an IP address from a private IP address space and also connected to **docker0 bridge**, as shown in the following diagram:

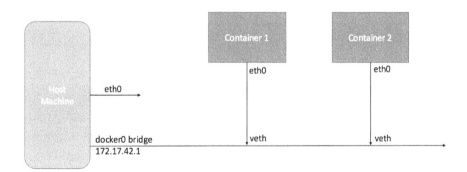

Two containers talking via Docker0 bridge

Both the preceding containers will be able to ping each other as well as reach the external world. For external access, their ports will be mapped to a host port. As mentioned in the previous section, containers use network namespaces. When the first container is created, a new network namespace is created for the container.

A **Virtual Ethernet (vEthernet** or **vEth)** link is created between the container and the Linux bridge. Traffic sent from the eth0 port of the container reaches the bridge through the vEth interface and gets switched thereafter:

```
# show linux bridges
$ sudo brctl show
```

The output of the preceding command will be similar to the following one with bridge name and the vEth interfaces on the containers it is mapped to:

```
$ bridge name  bridge              id    STP      enabled interfaces
  docker0      8000.56847afe9799 no      veth44cb727    veth98c3700
```

Connecting containers to the external world

The **iptables NAT** table on the host is used to masquerade all external connections, as shown here:

```
$ sudo iptables -t nat -L -n
. . .
Chain POSTROUTING (policy ACCEPT) target prot opt
source destination MASQUERADE all -- 172.17.0.0/16
!172.17.0.0/16
. . .
```

Reaching containers from the outside world

The port mapping is again done using the iptables NAT option in the host machine, as the following diagram shows, where port mapping of **Container 1** is done to communicate with the external world. We will look into it in detail in the later part of the chapter.

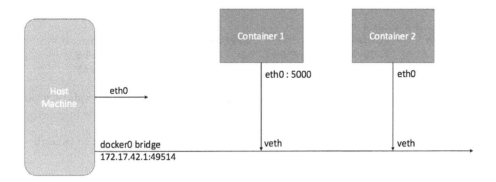

Port mapping of Container 1 to communicate with the external world

Docker server, by default, creates a `docker0` bridge inside the Linux kernel that can pass packets back and forth between other physical or virtual network interfaces so that they behave as a single ethernet network:

```
root@ubuntu:~# ifconfig
docker0    Link encap:Ethernet   HWaddr 56:84:7a:fe:97:99
           inet addr:172.17.42.1  Bcast:0.0.0.0  Mask:255.255.0.0
           inet6 addr: fe80::5484:7aff:fefe:9799/64 Scope:Link
           inet6 addr: fe80::1/64 Scope:Link
           . . .
           collisions:0 txqueuelen:0
           RX bytes:516868 (516.8 KB)   TX bytes:46460483 (46.4 MB)
eth0       Link encap:Ethernet   HWaddr 00:0c:29:0d:f4:2c
           inet addr:192.168.186.129  Bcast:192.168.186.255
Mask:255.255.255.0
```

Once we have one or more containers up and running, we can confirm that Docker has properly connected them to the docker0 bridge by running the `brctl` command on the host machine and looking at the interfaces column of the output. First, install the bridge utilities using the following command:

```
$ apt-get install bridge-utils
```

Here is a host with two different containers connected:

```
root@ubuntu:~# brctl show
bridge name     bridge id              STP enabled    interfaces
docker0         8000.56847afe9799      no             veth21b2e16
                                                       veth7092a45
```

Docker uses docker0 bridge settings whenever a container is created. It assigns a new IP address from the range available on the bridge whenever a new container is created:

```
root@ubuntu:~# docker run -t -i --name container1 ubuntu:latest
/bin/bash
root@e54e9312dc04:/# ifconfig
eth0 Link encap:Ethernet HWaddr 02:42:ac:11:00:07
inet addr:172.17.0.7 Bcast:0.0.0.0 Mask:255.255.0.0
inet6 addr: 2001:db8:1::242:ac11:7/64 Scope:Global
inet6 addr: fe80::42:acff:fe11:7/64 Scope:Link
UP BROADCAST RUNNING MULTICAST MTU:1500 Metric:1
...
root@e54e9312dc04:/# ip route
default via 172.17.42.1 dev eth0
172.17.0.0/16 dev eth0 proto kernel scope link src 172.17.0.7
```

By default, Docker provides a vnet docker0 that has the `172.17.42.1` IP address. Docker containers have IP addresses in the range of `172.17.0.0/16`
To change the default settings in Docker, modify the `/etc/default/docker` file.

Change the default bridge from `docker0` to `br0`:

```
# sudo service docker stop
# sudo ip link set dev docker0 down
# sudo brctl delbr docker0
# sudo iptables -t nat -F POSTROUTING
# echo 'DOCKER_OPTS="-b=br0"' >> /etc/default/docker
# sudo brctl addbr br0
# sudo ip addr add 192.168.10.1/24 dev br0
# sudo ip link set dev br0 up
# sudo service docker start
```

The following command displays the new bridge name and the IP address range of the Docker service:

```
root@ubuntu:~# ifconfig
br0        Link encap:Ethernet  HWaddr ae:b2:dc:ed:e6:af
           inet addr:192.168.10.1  Bcast:0.0.0.0  Mask:255.255.255.0
           inet6 addr: fe80::acb2:dcff:feed:e6af/64 Scope:Link
           UP BROADCAST RUNNING MULTICAST  MTU:1500  Metric:1
           RX packets:0 errors:0 dropped:0 overruns:0 frame:0
           TX packets:7 errors:0 dropped:0 overruns:0 carrier:0
           collisions:0 txqueuelen:0
           RX bytes:0 (0.0 B)  TX bytes:738 (738.0 B)
eth0       Link encap:Ethernet  HWaddr 00:0c:29:0d:f4:2c
           inet addr:192.168.186.129  Bcast:192.168.186.255
Mask:255.255.255.0
           inet6 addr: fe80::20c:29ff:fe0d:f42c/64 Scope:Link
              . . .
```

Configuring DNS

Docker provides hostname and DNS configuration for each container without building a custom image. It overlays the /etc files inside the container with virtual files where it can write new information.

This can be seen by running the mount command inside the container. Containers receive the same /resolv.conf as of the host machine when they are created initially. If a host's /resolv.conf file is modified, it will be reflected in the container's /resolv.conf file only when the container is restarted.

In Docker, you can set the dns options in two ways:

- Using docker run --dns=<ip-address>
- In the Docker daemon file, add DOCKER_OPTS="--dns ip-address"

> You can also specify the search domain using --dns-search=<DOMAIN>.

The following diagram shows the nameserver being configured in container using the DOCKER_OPTS setting in the Docker daemon file:

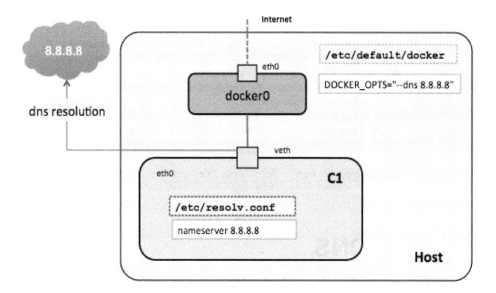

DOCKER_OPTS being used to set nameserver setting for Docker container

The main DNS files are as follows:

```
/etc/hostname
/etc/resolv.conf
/etc/hosts
```

The following is the command to add the DNS server:

```
# docker run --dns=8.8.8.8 --net="bridge" -t -i  ubuntu:latest
/bin/bash
```

Here's the command to add the hostname:

```
#docker run --dns=8.8.8.8 --hostname=docker-vm1  -t -i  ubuntu:latest
/bin/bash
```

Troubleshooting communication between containers and the external network

Packets can only pass between containers if the `ip_forward` parameter is set to 1. Usually, you will simply leave the Docker server at its default setting of `--ip-forward=true` and Docker will set `ip_forward` to 1 for you when the server starts up. To check the settings, use the following command:

```
# cat /proc/sys/net/ipv4/ip_forward
0
# echo 1 > /proc/sys/net/ipv4/ip_forward
# cat /proc/sys/net/ipv4/ip_forward
1
```

By enabling `ip-forward`, users can make communication between containers and the external world possible; it will also be needed for inter-container communication if you are in a multiple bridge setup:

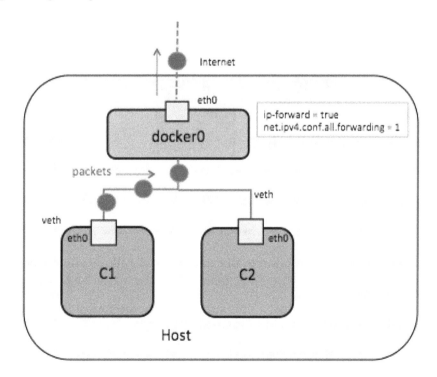

ip-forward = true forwards all the packets to/from the container to the external network

Docker will not delete or modify any pre-existing rules from Docker filter chain. This allows users to create rules to restrict access to containers. Docker uses docker0 bridge for packet flow between all containers in a single host. It adds a rule to the FORWARD chain in iptables (blank accept policy) for the packets to flow between two containers. The --icc=false option will DROP all the packets.

When the Docker daemon is configured with both --icc=false and --iptables=true and the Docker run is invoked with the --link= option, the Docker server will insert a pair of iptables ACCEPT rules for the new container to connect to the ports exposed by the other container–the ports that it mentioned in the EXPOSE lines of its Dockerfile:

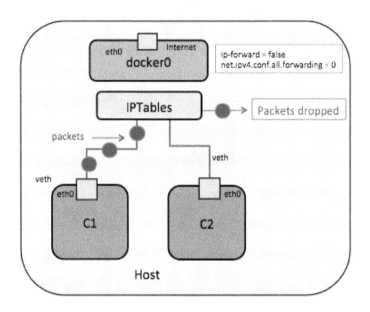

ip-forward = false forwards all the packets to/from the container to external network

By default, Docker's forward rule permits all external IPs. To allow only a specific IP or network to access the containers, insert a negated rule at the top of the Docker filter chain.

For example, you can restrict external access such that only the source IP 10.10.10.10 can access the containers using the following command:

```
#iptables -I DOCKER -i ext_if ! -s 10.10.10.10 -j DROP
```

References:
https://docs.docker.com/v1.5/articles/networking/
https://docs.docker.com/engine/userguide/networking/
http://containerops.org/

Restricting SSH access from one container to another

To restrict SSH access from one container to another, perform the following steps:

1. Create two containers, c1 and c2:

    ```
    # docker run -i -t --name c1 ubuntu:latest /bin/bash
    root@7bc2b6cb1025:/# ifconfig
    eth0 Link encap:Ethernet HWaddr 02:42:ac:11:00:05
     inet addr:172.17.0.5 Bcast:0.0.0.0 Mask:255.255.0.0
     inet6 addr: 2001:db8:1::242:ac11:5/64 Scope:Global
     inet6 addr: fe80::42:acff:fe11:5/64 Scope:Link
     ...
    # docker run -i -t --name c2 ubuntu:latest /bin/bash
    root@e58a9bf7120b:/# ifconfig
    eth0 Link encap:Ethernet HWaddr 02:42:ac:11:00:06
     inet addr:172.17.0.6 Bcast:0.0.0.0 Mask:255.255.0.0
     inet6 addr: 2001:db8:1::242:ac11:6/64 Scope:Global
     inet6 addr: fe80::42:acff:fe11:6/64 Scope:Link
    ```

2. We can test connectivity between the containers using the IP address we've just discovered. Let's see this now using the `ping` tool.

3. Let's go into the other container, c1, and try to ping c2:

    ```
    root@7bc2b6cb1025:/# ping 172.17.0.6
    PING 172.17.0.6 (172.17.0.6) 56(84) bytes of data.
    64 bytes from 172.17.0.6: icmp_seq=1 ttl=64 time=0.139 ms
    64 bytes from 172.17.0.6: icmp_seq=2 ttl=64 time=0.110 ms
    ^C
    --- 172.17.0.6 ping statistics ---
    2 packets transmitted, 2 received, 0% packet loss, time 999ms
    rtt min/avg/max/mdev = 0.110/0.124/0.139/0.018 ms
    root@7bc2b6cb1025:/#
    root@e58a9bf7120b:/# ping 172.17.0.5
    PING 172.17.0.5 (172.17.0.5) 56(84) bytes of data.
    64 bytes from 172.17.0.5: icmp_seq=1 ttl=64 time=0.270 ms
    64 bytes from 172.17.0.5: icmp_seq=2 ttl=64 time=0.107 ms
    ```

```
^C
--- 172.17.0.5 ping statistics ---

2 packets transmitted, 2 received, 0% packet loss, time 1002ms
rtt min/avg/max/mdev = 0.107/0.188/0.270/0.082 ms
root@e58a9bf7120b:/#
```

4. Install `openssh-server` on both the containers:

   ```
   #apt-get install openssh-server
   ```

5. Enable iptables on the host machine. Initially, you will be able to SSH from one container to another.

6. Stop the Docker service and add `DOCKER_OPTS="--icc=false --iptables=true"` in the `default docker` file of the host machine. This option will enable the iptables firewall and drop all ports between the containers. By default, iptables are not enabled on the host:

   ```
   root@ubuntu:~# iptables -L -n
   Chain INPUT (policy ACCEPT)
   target prot opt source destination
   Chain FORWARD (policy ACCEPT)
   target prot opt source destination
   DOCKER all -- 0.0.0.0/0 0.0.0.0/0
   ACCEPT all -- 0.0.0.0/0 0.0.0.0/0 ctstate RELATED,ESTABLISHED
   ACCEPT all -- 0.0.0.0/0 0.0.0.0/0
   DOCKER all -- 0.0.0.0/0 0.0.0.0/0
   ACCEPT all -- 0.0.0.0/0 0.0.0.0/0 ctstate RELATED,ESTABLISHED
   ACCEPT all -- 0.0.0.0/0 0.0.0.0/0
   ACCEPT all -- 0.0.0.0/0 0.0.0.0/0
   ACCEPT all -- 0.0.0.0/0 0.0.0.0/0

   #service docker stop
   #vi /etc/default/docker
   ```

7. The Docker Upstart and SysVinit configuration file, customize the location of the Docker binary (especially for development testing):

   ```
   #DOCKER="/usr/local/bin/docker"
   ```

8. Use `DOCKER_OPTS` to modify the daemon startup options:

```
#DOCKER_OPTS="--dns 8.8.8.8 --dns 8.8.4.4"
#DOCKER_OPTS="--icc=false --iptables=true"
```

9. Restart the Docker service:

```
# service docker start
```

10. Inspect the iptables:

```
root@ubuntu:~# iptables -L -n
Chain INPUT (policy ACCEPT)
target prot opt source destination
Chain FORWARD (policy ACCEPT)
target prot opt source destination
DOCKER all -- 0.0.0.0/0 0.0.0.0/0
ACCEPT all -- 0.0.0.0/0 0.0.0.0/0 ctstate RELATED, ESTABLISHED
ACCEPT all -- 0.0.0.0/0 0.0.0.0/0
DOCKER all -- 0.0.0.0/0 0.0.0.0/0
ACCEPT all -- 0.0.0.0/0 0.0.0.0/0 ctstate RELATED, ESTABLISHED
ACCEPT all -- 0.0.0.0/0 0.0.0.0/0
ACCEPT all -- 0.0.0.0/0 0.0.0.0/0
DROP all -- 0.0.0.0/0 0.0.0.0/0
```

The `DROP` rule has been added to the iptables of the host machine, which drops connection between the containers. Now you won't be able to SSH between the containers.

Linking containers

We can communicate or connect legacy containers using the `--link` parameter.

1. Create the first container that will act as the server–sshserver:

```
root@ubuntu:~# docker run -i -t -p 2222:22 --name sshserver ubuntu
bash
root@9770be5acbab:/#
Execute the iptables command and you can find a Docker chain rule
added.
#root@ubuntu:~# iptables -L -n
Chain INPUT (policy ACCEPT)
target     prot opt source              destination
Chain FORWARD (policy ACCEPT)
target     prot opt source              destination
Chain OUTPUT (policy ACCEPT)
target     prot opt source              destination
```

```
        Chain DOCKER (0 references)
        target     prot opt source          destination
        ACCEPT     tcp  --  0.0.0.0/0        172.17.0.3           tcp
dpt:22
```

2. Create a second container that acts like an SSH client:

```
root@ubuntu:~# docker run -i -t --name sshclient --link
sshserver:sshserver
ubuntu bash
root@979d46c5c6a5:/#
```

3. We can see that there are more rules added to the Docker chain rule:

```
root@ubuntu:~# iptables -L -n
Chain INPUT (policy ACCEPT)
target     prot opt source               destination
Chain FORWARD (policy ACCEPT)
target     prot opt source               destination
Chain OUTPUT (policy ACCEPT)
target     prot opt source               destination
Chain DOCKER (0 references)
target     prot opt source               destination
ACCEPT     tcp  --  0.0.0.0/0            172.17.0.3           tcp
dpt:22
ACCEPT     tcp  --  172.17.0.4           172.17.0.3           tcp
dpt:22
ACCEPT     tcp  --  172.17.0.3           172.17.0.4           tcp
spt:22
root@ubuntu:~#
```

The following diagram explains the communication between containers using the --link flag:

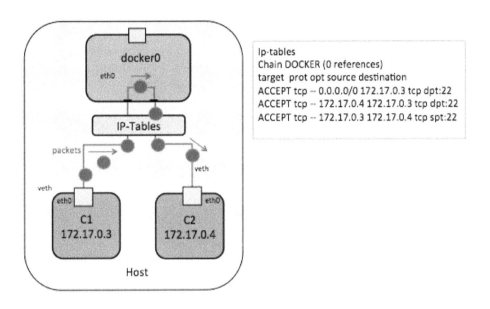

Ip-tables
Chain DOCKER (0 references)
target prot opt source destination
ACCEPT tcp -- 0.0.0.0/0 172.17.0.3 tcp dpt:22
ACCEPT tcp -- 172.17.0.4 172.17.0.3 tcp dpt:22
ACCEPT tcp -- 172.17.0.3 172.17.0.4 tcp spt:22

Docker–link creates private channels between containers

4. You can inspect your linked container with `docker inspect`:

```
root@ubuntu:~# docker inspect -f "{{ .HostConfig.Links }}"
sshclient
[/sshserver:/sshclient/sshserver]
```

5. Now you can successfully SSH into the SSH server with its IP:

```
#ssh root@172.17.0.3 -p 22
```

Using the --link parameter, Docker creates a secure channel between the containers that doesn't need to expose any ports externally on the containers.

libnetwork and the Container Network Model

libnetwork is implemented in Go for connecting Docker containers. The aim is to provide a **Container Network Model** (**CNM**) that helps programmers provide the abstraction of network libraries. The long-term goal of libnetwork is to follow the Docker and Linux philosophy to deliver modules that work independently. libnetwork has the aim for providing a composable need for networking in containers. It also aims to modularize the networking logic in the Docker Engine and libcontainer to a single, reusable library by doing the following things:

- Replacing the networking module of the Docker Engine with libnetwork
- Allowing local and remote drivers to provide networking to containers
- Providing a `dnet` tool for managing and testing libnetwork–however, this is still a work in progress

Reference:
`https://github.com/docker/libnetwork/issues/45`

libnetwork implements the CNM. It formalizes the steps required to provide networking for containers while providing an abstraction that can be used to support multiple network drivers. Its endpoint APIs are primarily used for managing the corresponding object and bookkeeping them in order to provide the level of abstraction as required by the CNM.

CNM objects

The CNM is built on three main components, as shown in the following diagram:

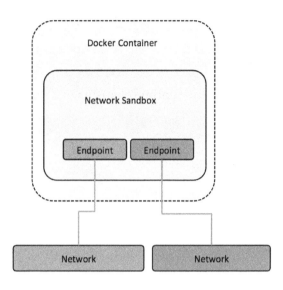

Network sandbox model of libnetwork

Reference:
https://www.docker.com

Sandbox

A sandbox contains the configuration of a container's network stack that includes management of routing tables, the container's interface, and DNS settings. The implementation of a sandbox can be a Linux network namespace, a FreeBSD jail, or another similar concept.

A sandbox may contain many endpoints from multiple networks. It also represents a container's network configuration, such as IP address, MAC address, and DNS entries.

libnetwork makes use of the OS-specific parameters to populate the network configuration represented by a sandbox. It provides a framework to implement a sandbox in multiple OSes.

Netlink is used to manage the routing table in namespace and currently two implementations of a sandbox exist–namespace_linux.go and configure_linux.go–to uniquely identify the path on the host filesystem. A sandbox is associated with a single Docker container.

The following data structure shows the runtime elements of a sandbox:

```
type sandbox struct {
        id             string
         containerID   string
        config         containerConfig
        osSbox         osl.Sandbox
        controller     *controller
        refCnt         int
        endpoints      epHeap
        epPriority     map[string]int
        joinLeaveDone  chan struct{}
        dbIndex        uint64
        dbExists       bool
        isStub         bool
        inDelete       bool
        sync.Mutex
}
```

A new sandbox is instantiated from a network controller (which is explained in detail later):

```
func (c *controller) NewSandbox(containerID string, options
...SandboxOption)
   (Sandbox, error) {
      .....
}
```

Endpoint

An endpoint joins a sandbox to a network and provides connectivity for services exposed by a container to the other containers deployed in the same network. It can be an internal port of Open vSwitch or a similar vEth pair.

An endpoint can belong to only one network and may only belong to one sandbox. It represents a service and provides various APIs to create and manage the endpoint. It has a global scope but gets attached to only one network.

An endpoint is specified by the following struct:

```
type endpoint struct {
    name          string
    id            string
    network       *network
    iface         *endpointInterface
    joinInfo      *endpointJoinInfo
    sandboxID     string
    exposedPorts  []types.TransportPort
    anonymous     bool
    generic       map[string]interface{}
    joinLeaveDone chan struct{}
    prefAddress   net.IP
    prefAddressV6 net.IP
    ipamOptions   map[string]string
    dbIndex       uint64
    dbExists      bool
    sync.Mutex
}
```

An endpoint is associated with a unique ID and name. It is attached to a network and a sandbox ID. It is also associated with a IPv4 and IPv6 address spaces. Each endpoint is associated with an endpoint interface.

Network

A group of endpoints that are able to communicate with each other directly is called a **network**. It provides the required connectivity within the same host or multiple hosts and whenever a network is created or updated, the corresponding driver is notified. An example is a VLAN or Linux bridge that has a global scope within a cluster.

A networks are controlled from a network controller, which we will discuss in the next section. Every network has a name, address space, ID, and network type:

```
type network struct {
    ctrlr        *controller
    name         string
    networkType  string
    id           string
    ipamType     string
    addrSpace    string
```

```
        ipamV4Config  []*IpamConf
        ipamV6Config  []*IpamConf
        ipamV4Info    []*IpamInfo
        ipamV6Info    []*IpamInfo
        enableIPv6    bool
        postIPv6      bool
        epCnt         *endpointCnt
        generic       options.Generic
        dbIndex       uint64
        svcRecords    svcMap
        dbExists      bool
        persist       bool
        stopWatchCh   chan struct{}
        drvOnce       *sync.Once
        internal      bool
        sync.Mutex
}
```

Network controller

A network controller object provides APIs to create and manage a network object. It is an entry point to the libnetwork by binding a particular driver to a given network, and it supports multiple active drivers, both inbuilt and remote. A network controller allows users to bind a particular driver to a given network:

```
type controller struct {
    id              string
    drivers         driverTable
    ipamDrivers     ipamTable
    sandboxes       sandboxTable
    cfg             *config.Config
    stores          []datastore.DataStore
    discovery       hostdiscovery.HostDiscovery
    extKeyListener  net.Listener
    watchCh         chan *endpoint
    unWatchCh       chan *endpoint
    svcDb           map[string]svcMap
    nmap            map[string]*netWatch
    defOsSbox       osl.Sandbox
    sboxOnce        sync.Once
    sync.Mutex
}
```

Each network controller has reference to the following things:

- One or more drivers in a data structure driver table
- One or more sandboxes in a data structure
- A data store
- An ipamTable

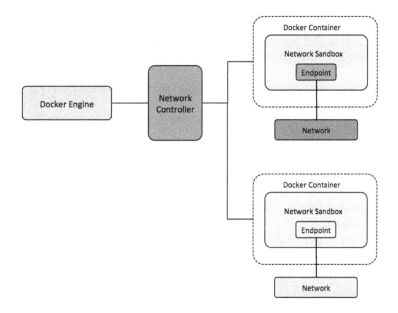

Network controller handling the network between Docker container and Docker engine

The preceding diagram shows how the network controller sits between the Docker Engine, containers, and the networks they are attached to.

CNM attributes

The following are the CNM attributes:

- **Options:** These are not end user visible but are the key-value pairs of data to provide a flexible mechanism to pass driver-specific configuration from user to driver directly. libnetwork operates on the options only if a key matches a well-known label and as a result of this a value is picked up that is represented by a generic object.

- **Labels:** These are a subset of options that are end user variable represented in the UI using the `--labels` option. Their main function is to perform driver-specific operations, and they are passed from the UI.

CNM life cycle

Consumers of the CNM interact through the CNM objects and its APIs to network the containers that they manage; drivers register with a network controller.

Built-in drivers are registered inside libnetwork, while remote drivers register with libnetwork using a plugin mechanism.

Each driver handles a particular network type as explained as follows:

- A network controller object is created using the `libnetwork.New()` API to manage the allocation of networks and optionally configure a driver with driver-specific options. The network object is created using the controller's `NewNetwork()` API, a `name`, and a `NetworkType` is added as a parameter.
- The `NetworkType` parameter helps to choose a driver and binds the created network to that driver. All operations on network will be handled by the driver that is created using the preceding API.
- The `Controller.NewNetwork()` API takes in an optional options parameter that carries driver-specific options and labels, which the driver can use for its purpose.
- `Network.CreateEndpoint()` is called to create a new endpoint in a given network. This API also accepts optional options parameters that vary with the driver.
- `CreateEndpoint()` can choose to reserve IPv4/IPv6 addresses when an endpoint is created in a network. The driver assigns these addresses using the `InterfaceInfo` interface defined in `driverapi`. The IPv4/IPv6 addresses are needed to complete the endpoint as service definition along with the ports that the endpoint exposes. A service endpoint is a network address and the port number that the application container is listening on.
- `Endpoint.Join()` is used to attach a container to an endpoint. The `Join` operation will create a sandbox, if one doesn't exist for that container. The drivers make use of the sandbox key to identify multiple endpoints attached to the same container.

There is a separate API to create an endpoint and another to join the endpoint.

An endpoint represents a service that is independent of the container. When an endpoint is created, it has resources reserved for a container to get attached to the endpoint later. It gives a consistent networking behavior.

- `Endpoint.Leave()` is invoked when a container is stopped. The driver can clean up the states that it allocated during the `Join()` call. libnetwork deletes the sandbox when the last referencing endpoint leaves the network.
- libnetwork keeps holding on to IP addresses as long as the endpoint is still present. These will be reused when the container (or any container) joins again. It ensures that the container's resources are reused when they are stopped and started again.
- `Endpoint.Delete()` deletes an endpoint from a network. This results in deleting the endpoint and cleaning up the cached `sandbox.Info`.
- `Network.Delete()` is used to delete a network. Deleting is allowed if there are no endpoints attached to the network.

Docker networking tools based on overlay and underlay networks

An overlay is a virtual network that is built on top of anunderlying network infrastructure (the underlay). The purpose is to implement a network service that is not available in the physical network.

Network overlay dramatically increases the number of virtual subnets that can be created on top of the physical network, which in turn supports multi-tenancy and virtualization features.

Every container in Docker is assigned with an IP address that is used for communication with other containers. If a container has to communicate to the external network, you set up networking in the host system and expose or map the port from the container to the host machine. With this application running inside, containers will not be able to advertise their external IP and ports as the information is not available to them.

The solution is to somehow assign unique IPs to each Docker container across all hosts and have some networking product that routes traffic between the hosts.

There are different projects and tools to help with Docker networking, as follows:

- Flannel
- Weave
- Project Calico

Flannel

Flannel gives each container an IP that can be used for container-to-container communication. By packet encapsulation, it creates a virtual overlay network over host network. By default, flannel provides a /24 subnet to the hosts, from which the Docker daemon will allocate IPs to the containers.

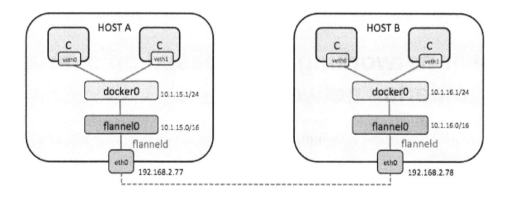

Communication between containers using Flannel

Flannel runs an agent, flanneld, on each host and is responsible for allocating a subnet lease out of a preconfigured address space. Flannel uses etcd (https://github.com/coreos/etcd) to store the network configuration, allocated subnets, and auxiliary data (such as the host's IP).

In order to provide encapsulation, Flannel uses the **Universal TUN/TAP** device and creates an overlay network using UDP to encapsulate IP packets. The subnet allocation is done with the help of etcd, which maintains the overlay subnet to host mappings.

Weave

Weave creates a virtual network that connects Docker containers deployed across hosts/VMs and enables their automatic discovery.

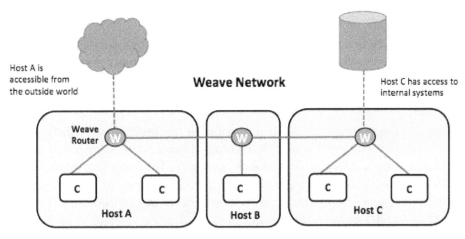

Weave Network

Weave can traverse firewalls and operate in partially connected networks. Traffic can be optionally encrypted, allowing hosts/VMs to be connected across an untrusted network.

Weave augments Docker's existing (single host) networking capabilities, such as the docker0 bridge, so that these can continue to be used by the containers.

Project Calico

Project Calico provides a scalable networking solution for connecting containers, VMs, or bare metal. Calico provides connectivity using the scalable IP networking principle as a layer 3 approach. Calico can be deployed without overlays or encapsulation. The Calico service should be deployed as a container on each node. It provides each container with its own IP address and, also, handles all the necessary IP routing, security policy rules, and distribution of routes across a cluster of nodes.

The Calico architecture contains four important components in order to provide better networking solutions:

- **Felix**, the Calico worker process, is the heart of the Calico networking that primarily routes and provides the desired connectivity to and from the workloads on the host. It also provides the interface to the kernel for outgoing endpoint traff

- **BIRD**, the route ic. BIRD, the route distribution open source BGP, exchanges routing information between hosts. The kernel endpoints that are picked up by BIRD are distributed to BGP peers in order to provide inter-host routing. Two BIRD processes run in the *calico-node* container, IPv4 (bird) and one for IPv6 (bird6)

- **confd**, a templating process to autogenerate configuration for BIRD, monitors the `etcd` store for any changes to BGP configuration, such as log levels and IPAM information. `confd` also dynamically generates BIRD configuration files based on data from `etcd` and is trigger automatically as updates are applied to the data. `confd` triggers BIRD to load new files whenever the configuration file is changed.

- **calicoctl** is the command line used to configure and start the Calico service. It even allows the data store (`etcd`) to define and apply security policy. The tool also provides the simple interface for general management of Calico configuration irrespective of whether Calico is running on VMs, containers, or bare metal. The following commands are supported at `calicoctl`;

```
$ calicoctl
Override the host:port of the ETCD server by setting the
  environment
variable
ETCD_AUTHORITY [default: 127.0.0.1:2379]
Usage: calicoctl <command> [<args>...]
status          Print current status information
node            Configure the main calico/node container and
  establish
                Calico networking
container       Configure containers and their addresses
profile         Configure endpoint profiles
endpoint        Configure the endpoints assigned to existing
  containers
pool            Configure ip-pools
bgp             Configure global bgp
ipam            Configure IP address management
checksystem     Check for incompatibilities on the host
  system
diags           Save diagnostic information
version         Display the version of calicoctl
```

```
config             Configure low-level component configuration
See 'calicoctl <command> --help' to read about a specific
subcommand.
```

As per the official GitHub page of the Calico repository
(`https://github.com/projectcalico/calico-containers`), the following integration of
Calico exists:

- Calico as a Docker network plugin
- Calico without Docker networking
- Calico with Kubernetes
- Calico with Mesos
- Calico with Docker Swarm

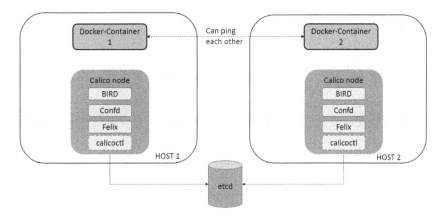

Calico Architecture

Configuring an overlay network with the Docker Engine swarm node

With the release of Docker 1.9, the multi-host and overlay network has become one of
its primary feature. It enables private networks that can be established to connect multiple
containers. We will be creating the overlay network on a manager node running in swarm
cluster without an external key-value store. The swarm network will make the network
available to the nodes in the swarm that require it for a service.

When we deploy the service that uses overlay network, the manager automatically extends the network to the nodes that are running the service tasks. Multi-host networking requires a store for service discovery, so now we will create a Docker machine to run this service.

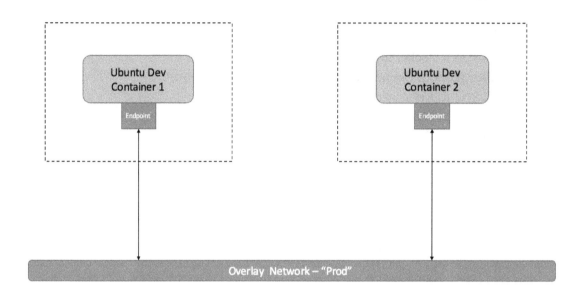

Overlay network across multiple hosts

For the following deployment, we will be using Docker machine application that creates the Docker daemon on a virtualization or cloud platform. For the virtualization platform, we will be using VMware fusion as the provider.

The Docker-machine installation is as follows:

```
$ curl -L https://github.com/docker/machine/releases/download/
v0.7.0/docker-machine-`uname -s`-`uname -m` > /usr/local/bin/
docker-machine && \
> chmod +x /usr/local/bin/docker-machine
     % Total    % Received % Xferd  Average Speed   Time    Time     Time
Current
                                   Dload  Upload   Total   Spent    Left
Speed
    100    601    0    601    0      0    266      0 --:--:--  0:00:02 --:--:-
-    266
    100 38.8M 100 38.8M    0      0   1420k      0  0:00:28  0:00:28 --:--:-
```

```
- 1989k

    $ docker-machine version
    docker-machine version 0.7.0, build a650a40
```

Mulithost networking requires a store for service discovery, so we will create a Docker machine to run that service, creating the new Docker daemon:

```
    $ docker-machine create \
    >    -d vmwarefusion \
    >    swarm-consul
    Running pre-create checks...
    (swarm-consul) Default Boot2Docker ISO is out-of-date, downloading the
latest
    release...
    (swarm-consul) Latest release for github.com/boot2docker/boot2docker is
    v1.12.1
    (swarm-consul) Downloading
    ...
```

> To see how to connect your Docker client to the Docker Engine running on this virtual machine, run
> `docker-machine env swarm-consul`.

We'll start the consul container for service discovery:

```
    $(docker-machine config swarm-consul) run \
    >        -d \
    >        --restart=always \
    >        -p "8500:8500" \
    >        -h "consul" \
    >        progrium/consul -server -bootstrap
    Unable to find image 'progrium/consul:latest' locally
    latest: Pulling from progrium/consul
    ...
    Digest:
    sha256:8cc8023462905929df9a79ff67ee435a36848ce7a10f18d6d0faba9306b97274
    Status: Downloaded newer image for progrium/consul:latest
    d482c88d6a1ab3792aa4d6a3eb5e304733ff4d622956f40d6c792610ea3ed312
```

Create two Docker daemons to run the Docker cluster, the first daemon is the swarm node that will automatically run a Swarm container used to coordinate the cluster:

```
    $ docker-machine create \
    >    -d vmwarefusion \
    >    --swarm \
```

```
>    --swarm-master \
>    --swarm-discovery="consul://$(docker-machine ip swarm-
 consul):8500" \
>    --engine-opt="cluster-store=consul://$(docker-machine ip swarm-
consul):8500" \
>    --engine-opt="cluster-advertise=eth0:2376" \
>    swarm-0
Running pre-create checks...
Creating machine...
(swarm-0) Copying
 /Users/vkohli/.docker/machine/cache/boot2docker.iso to
/Users/vkohli/.docker/machine/machines/swarm-0/boot2docker.iso...
(swarm-0) Creating SSH key...
(swarm-0) Creating VM...
...
```

Docker is up and running!

> To see how to connect your Docker client to the Docker Engine running on this virtual machine, run `docker-machine env swarm-0`.

The second daemon is the Swarm `secondary` node that will automatically run a Swarm container and report the state back to the `master` node:

```
$ docker-machine create \
>    -d vmwarefusion \
>    --swarm \
>    --swarm-discovery="consul://$(docker-machine ip swarm-
 consul):8500" \
>    --engine-opt="cluster-store=consul://$(docker-machine ip swarm-
consul):8500" \
>    --engine-opt="cluster-advertise=eth0:2376" \
>    swarm-1
Running pre-create checks...
Creating machine...
(swarm-1) Copying
 /Users/vkohli/.docker/machine/cache/boot2docker.iso to
/Users/vkohli/.docker/machine/machines/swarm-1/boot2docker.iso...
(swarm-1) Creating SSH key...
(swarm-1) Creating VM...
...
```

Docker is up and running!

To see how to connect your Docker client to the Docker Engine running on this virtual machine, run `docker-machine env swarm-1`.

Docker executable will communicate with one Docker daemon. Since we are in a cluster, we'll ensure the communication of the Docker daemon to the cluster by running the following command:

```
$ eval $(docker-machine env --swarm swarm-0)
```

After this, we'll create a private `prod` network with an overlay driver:

```
$ docker $(docker-machine config swarm-0) network create --driver
overlay prod
```

We will be starting the two dummy `ubuntu:12.04` containers using the `--net` parameter:

```
$ docker run -d -it --net prod --name dev-vm-1 ubuntu:12.04
426f39dbcb87b35c977706c3484bee20ae3296ec83100926160a39190451e57a
```

In the following code snippet, we can see that this Docker container has two network interfaces: one connected to the private overlay network and another to the Docker bridge:

```
$ docker attach 426
root@426f39dbcb87:/# ip address
23: eth0@if24: <BROADCAST,MULTICAST,UP,LOWER_UP> mtu 1450 qdisc
 noqueue state
UP
    link/ether 02:42:0a:00:00:02 brd ff:ff:ff:ff:ff:ff
    inet 10.0.0.2/24 scope global eth0
       valid_lft forever preferred_lft forever
    inet6 fe80::42:aff:fe00:2/64 scope link
       valid_lft forever preferred_lft forever
25: eth1@if26: <BROADCAST,MULTICAST,UP,LOWER_UP> mtu 1500 qdisc
 noqueue state
UP
    link/ether 02:42:ac:12:00:02 brd ff:ff:ff:ff:ff:ff
    inet 172.18.0.2/16 scope global eth1
       valid_lft forever preferred_lft forever
    inet6 fe80::42:acff:fe12:2/64 scope link
       valid_lft forever preferred_lft forever
```

The other container will also be connected to the `prod` network interface existing on the other host:

```
$ docker run -d -it --net prod --name dev-vm-7 ubuntu:12.04
d073f52a7eaacc0e0cb925b65abffd17a588e6178c87183ae5e35b98b36c0c25
$ docker attach d073
root@d073f52a7eaa:/# ip address
26: eth0@if27: <BROADCAST,MULTICAST,UP,LOWER_UP> mtu 1450 qdisc
 noqueue state
UP
    link/ether 02:42:0a:00:00:03 brd ff:ff:ff:ff:ff:ff
    inet 10.0.0.3/24 scope global eth0
       valid_lft forever preferred_lft forever
    inet6 fe80::42:aff:fe00:3/64 scope link
       valid_lft forever preferred_lft forever
28: eth1@if29: <BROADCAST,MULTICAST,UP,LOWER_UP> mtu 1500 qdisc
 noqueue state
UP
    link/ether 02:42:ac:12:00:02 brd ff:ff:ff:ff:ff:ff
    inet 172.18.0.2/16 scope global eth1
       valid_lft forever preferred_lft forever
    inet6 fe80::42:acff:fe12:2/64 scope link
       valid_lft forever preferred_lft forever
root@d073f52a7eaa:/#
```

This is how a private network can be configured across hosts in the Docker Swarm cluster.

Comparison of all multi-host Docker networking solutions

	Calico	Flannel	Weave	Docker Overlay N/W
Network Model	Layer-3 Solution	VxLAN or UDP	VxLAN or UDP	VxLAN
Name Service	No	No	Yes	No
Protocol Support	TCP,UDP, ICMP & ICMPv6	All	All	All
Distributed Storage	Yes	Yes	No	Yes
Encryption Channel	No	TLS	NaCI Library	No

Configuring OpenvSwitch (OVS) to work with Docker

Open vSwitch (**OVS**) is an open source **OpenFlow** capable virtual switch that is typically used with hypervisors to interconnect virtual machines within a host and between different hosts across networks. Overlay networks need to create a virtual data path using supported tunneling encapsulations, such as VXLAN or GRE.

The overlay data path is provisioned between tunnel endpoints residing in the Docker host that gives the appearance of all hosts within a given provider segment being directly connected to one another.

As a new container comes online, the prefix is updated in the routing protocol announcing its location via a tunnel endpoint. As the other Docker hosts receive the updates, the forwarding is installed into OVS for which tunnel endpoint the host resides. When the host is deprovisioned, a similar process occurs and tunnel endpoint Docker hosts remove the forwarding entry for the deprovisioned container:

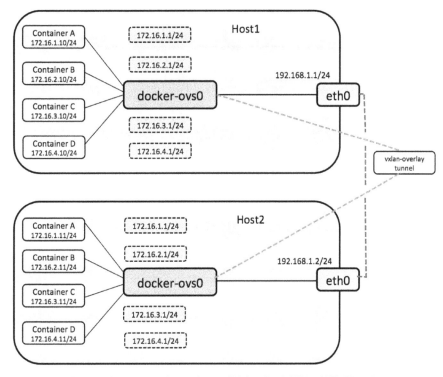

Communication between containers running on multiple hosts through OVS based VXLAN tunnels

By default, Docker uses the Linux docker0 bridge; however, there are cases where OVS might be required instead of the Linux bridge. A single Linux bridge can handle 1,024 ports only; this limits the scalability of Docker as we can only create 1,024 containers, each with a single network interface.

Troubleshooting OVS single host setup

Install OVS on a single host, create two containers, and connect them to an OVS bridge:

1. Install OVS:

   ```
   $ sudo apt-get install openvswitch-switch
   ```

2. Install the ovs-docker utility:

   ```
   $ cd /usr/bin
   $ wget https://raw.githubusercontent.com/openvswitch/ovs/master
   /utilities/ovs-docker
   $ chmod a+rwx ovs-docker
   ```

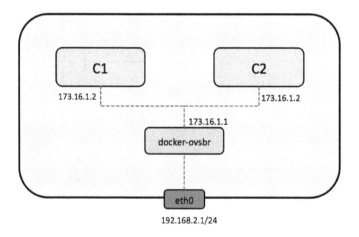

Single host OVS

3. Create an OVS bridge.

4. Here, we will be adding a new OVS bridge and configuring it so that we can get the containers connected on a different network:

```
$ ovs-vsctl add-br ovs-br1
$ ifconfig ovs-br1 173.16.1.1 netmask 255.255.255.0 up
```

5. Add a port from the OVS bridge to the Docker container.

6. Create two ubuntu Docker containers:

```
$ docker run -i-t --name container1 ubuntu /bin/bash
$ docker run -i-t --name container2 ubuntu /bin/bash
```

7. Connect the container to the OVS bridge:

```
# ovs-docker add-port ovs-br1 eth1 container1 --
  ipaddress=173.16.1.2/24
# ovs-docker add-port ovs-br1 eth1 container2 --
  ipaddress=173.16.1.3/24
```

8. Test the connection between the two containers connected using the OVS bridge with the ping command. First, find out their IP addresses:

```
# docker exec container1 ifconfig
eth0      Link encap:Ethernet  HWaddr 02:42:ac:10:11:02
inet addr:172.16.17.2  Bcast:0.0.0.0  Mask:255.255.255.0
inet6 addr: fe80::42:acff:fe10:1102/64 Scope:Link
...
# docker exec container2 ifconfig
eth0      Link encap:Ethernet  HWaddr 02:42:ac:10:11:03
inet addr:172.16.17.3  Bcast:0.0.0.0  Mask:255.255.255.0
inet6 addr: fe80::42:acff:fe10:1103/64 Scope:Link
...
```

9. Since we know the IP Address of container1 and container2, we can run the following command:

```
# docker exec container2 ping 172.16.17.2
PING 172.16.17.2 (172.16.17.2) 56(84) bytes of data.
64 bytes from 172.16.17.2: icmp_seq=1 ttl=64 time=0.257 ms
64 bytes from 172.16.17.2: icmp_seq=2 ttl=64 time=0.048 ms
64 bytes from 172.16.17.2: icmp_seq=3 ttl=64 time=0.052 ms
# docker exec container1 ping 172.16.17.2
PING 172.16.17.2 (172.16.17.2) 56(84) bytes of data.
64 bytes from 172.16.17.2: icmp_seq=1 ttl=64 time=0.060 ms
64 bytes from 172.16.17.2: icmp_seq=2 ttl=64 time=0.035 ms
64 bytes from 172.16.17.2: icmp_seq=3 ttl=64 time=0.031 ms
```

Troubleshooting OVS multiple host setups

First, we will connect Docker containers on multiple hosts using OVS:

Let us consider our setup, as shown in the following diagram, that contains two hosts–Host1 and Host2—running Ubuntu 14.04:

1. Install Docker and OVS on both the hosts:

   ```
   # wget -qO- https://get.docker.com/ | sh
   # sudo apt-get install openvswitch-switch
   ```

2. Install the ovs-docker utility:

   ```
   # cd /usr/bin
   # wget https://raw.githubusercontent.com/openvswitch/ovs
   /master/utilities/ovs-docker
   # chmod a+rwx ovs-docker
   ```

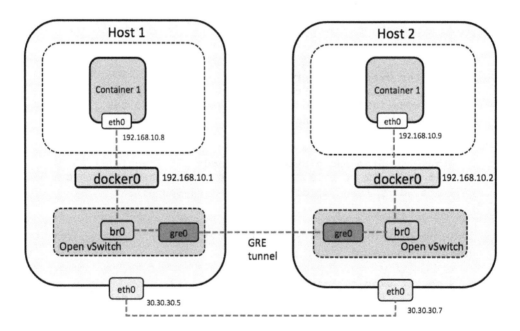

Multi Host container communication with OVS

3. Docker chooses a random network to run its containers by default. It creates a docker0 bridge and assigns an IP address (172.17.42.1) to it. So, both the Host1 and Host2 docker0 bridge IP addresses are the same, due to which it is difficult for containers in both the hosts to communicate. To overcome this, let's assign static IP addresses to the network, that is, (192.168.10.0/24).

To change the default Docker subnet:

1. Execute the following commands on Host1:

```
$ service docker stop
$ ip link set dev docker0 down
$ ip addr del 172.17.42.1/16 dev docker0
$ ip addr add 192.168.10.1/24 dev docker0
$ ip link set dev docker0 up
$ ip addr show docker0
$ service docker start
```

2. Add the br0 OVS bridge:

```
$ ovs-vsctl add-br br0
```

3. Create the tunnel to the other host:

```
$ ovs-vsctl add-port br0 gre0 -- set interface gre0 type=gre
options:remote_ip=30.30.30.8
```

4. Add the br0 bridge to the docker0 bridge:

```
$ brctl addif docker0 br0
```

5. Execute the following commands on Host2:

```
$ service docker stop
$ iptables -t nat -F POSTROUTING
$ ip link set dev docker0 down
$ ip addr del 172.17.42.1/16 dev docker0
$ ip addr add 192.168.10.2/24 dev docker0
$ ip link set dev docker0 up
$ ip addr show docker0
$ service docker start
```

6. Add the `br0` OVS bridge:

```
$ ip link set br0 up
$ ovs-vsctl add-br br0
```

7. Create the tunnel to the other host and attach it to the:

```
# br0 bridge
$ ovs-vsctl add-port br0 gre0 -- set interface gre0 type=gre
options:remote_ip=30.30.30.7
```

8. Add the `br0` bridge to the `docker0` bridge:

```
$ brctl addif docker0 br0
```

The docker0 bridge is attached to another bridge–`br0`. This time, it's an OVS bridge, which means that all traffic between the containers is routed through `br0` too. Additionally, we need to connect together the networks from both the hosts in which the containers are running. A GRE tunnel (`http://en.wikipedia.org/w iki/Generic_Routing_Encapsulation`) is used for this purpose. This tunnel is attached to the `br0` OVS bridge and, as a result, to `docker0` as well. After executing the preceding commands on both the hosts, you should be able to ping the `docker0` bridge addresses from both the hosts.

On Host1:

```
$ ping 192.168.10.2
PING 192.168.10.2 (192.168.10.2) 56(84) bytes of data.
64 bytes from 192.168.10.2: icmp_seq=1 ttl=64 time=0.088 ms
64 bytes from 192.168.10.2: icmp_seq=2 ttl=64 time=0.032 ms
^C
--- 192.168.10.2 ping statistics ---
2 packets transmitted, 2 received, 0% packet loss, time 999ms
rtt min/avg/max/mdev = 0.032/0.060/0.088/0.028 ms
```

On Host2:

```
$ ping 192.168.10.1
PING 192.168.10.1 (192.168.10.1) 56(84) bytes of data.
64 bytes from 192.168.10.1: icmp_seq=1 ttl=64 time=0.088 ms
64 bytes from 192.168.10.1: icmp_seq=2 ttl=64 time=0.032 ms
^C
--- 192.168.10.1 ping statistics ---
2 packets transmitted, 2 received, 0% packet loss, time 999ms
rtt min/avg/max/mdev = 0.032/0.060/0.088/0.028 ms
```

9. Create containers on both the hosts.

   On Host1, use the following command:

   ```
   $ docker run -t -i --name container1 ubuntu:latest /bin/bash
   ```

   On Host2, use the following command:

   ```
   $ docker run -t -i --name container2 ubuntu:latest /bin/bash
   ```

Now we can ping `container2` from `container1`. In this way, we connect Docker containers on multiple hosts using OVS.

Summary

In this chapter, we learnt how Docker networking is powered with docker0 bridge, its troubleshooting issues, and configuration. We also looked at troubleshooting the communication issues between Docker networks and the external network. Following that, we did some deep dive into libnetwork and the CNM and its life cycle. Then, we looked into containers' communication across multiple hosts using different networking options, such as Weave, OVS, Flannel, and Docker's latest overlay network, with comparison, and the troubleshooting issues involved in their configuration.

We saw that Weave creates a virtual network, OVS uses GRE tunneling technology, Flannel provides a separate subnet, and the Docker overlay sets up each host to connect containers on multiple hosts. After that, we looked into Docker network configuration with OVS and troubleshooting the single host and multiple host setup.

8
Managing Docker Containers with Kubernetes

In the previous chapter, we learned about Docker networking and how to troubleshoot networking issues. In this chapter, we will introduce Kubernetes.

Kubernetes is a container-cluster management tool. Currently, it supports Docker and Rocket. It is an open-sourced project by Google and it was launched in June 2014 at Google I/O. It supports deployment on various cloud providers, such as GCE, Azure, AWS, vSphere, and Bare Metal. The Kubernetes manager is lean, portable, extensible, and self-healing.

In this chapter, we will cover the following:

- An introduction to Kubernetes
- Deploying Kubernetes on Bare Metal
- Deploying Kubernetes on Minikube
- Deploying Kubernetes on AWS and vSphere
- Deploying a pod
- Deploying Kubernetes in a production environment
- Debugging Kubernetes issues

Kubernetes has various important components, as follows:

- **Node**: This is a physical or virtual machine that is part of a Kubernetes cluster, running the Kubernetes and Docker services, onto which pods can be scheduled.
- **Master**: This maintains the runtime state of Kubernetes' server runtime. It is the point of entry for all the client calls to configure and manage Kubernetes components.

- **Kubectl**: This is the command-line tool used to interact with the Kubernetes cluster to provide master access to Kubernetes APIs. Through it, the user can deploy, delete, and list pods.
- **Pod**: This is the smallest scheduling unit in Kubernetes. It is a collection of Docker containers that share volumes and don't have port conflicts. It can be created by defining a simple JSON file.
- **Replication controller**: This manages the lifecycle of the pod and ensures that the specified number of pods are running at any given time by creating or killing pods as required.
- **Label**: Labels are used to identify and organize pods and services based on key-value pairs:

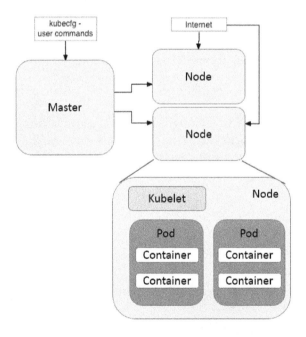

Kubernetes master/minion flow

Deploying Kubernetes on Bare Metal machine

Kubernetes can be deployed on the Bare Metal Fedora or Ubuntu machines. Even the Fedora and Ubuntu virtual machine can be deployed in vSphere, workstation, or VirtualBox. For the following tutorial, we'll be looking at Kubernetes deployment on a single Fedora 24 machine, which will be acting as master, as well as node to deploy `k8s` pods:

1. Enable the Kubernetes testing YUM repository:

   ```
   yum -y install --enablerepo=updates-testing kubernetes
   ```

2. Install `etcd` and `iptables-services`:

   ```
   yum -y install etcd iptables-services
   ```

3. In `/etcd/hosts`, set the Fedora master and Fedora node:

   ```
   echo "192.168.121.9  fed-master
   192.168.121.65  fed-node" >> /etc/hosts
   ```

4. Disable the firewall and `iptables-services`:

   ```
   systemctl disable iptables-services firewalld
   systemctl stop iptables-services firewalld
   ```

5. Edit the `/etcd/kubernetes/config` file:

   ```
   # Comma separated list of nodes in the etcd cluster
   KUBE_MASTER="--master=http://fed-master:8080"
   # logging to stderr means we get it in the systemd journal
   KUBE_LOGTOSTDERR="--logtostderr=true"
   # journal message level, 0 is debug
   KUBE_LOG_LEVEL="--v=0"
   # Should this cluster be allowed to run privileged docker
   containers
   KUBE_ALLOW_PRIV="--allow-privileged=false"
   ```

6. Edit the contents of the `/etc/kubernetes/apiserver` file:

```
# The address on the local server to listen to.
KUBE_API_ADDRESS="--address=0.0.0.0"

# Comma separated list of nodes in the etcd cluster
KUBE_ETCD_SERVERS="--etcd-servers=http://127.0.0.1:2379"

# Address range to use for services
KUBE_SERVICE_ADDRESSES="--service-cluster-ip-
range=10.254.0.0/16"

# Add your own!
KUBE_API_ARGS=""
```

7. The `/etc/etcd/etcd.conf` file should have the following line uncommented in order to listen on port 2379, as Fedora 24 uses etcd 2.0:

```
ETCD_LISTEN_CLIENT_URLS="http://0.0.0.0:2379"
```

8. **The Kubernetes node setup can be done on separate hosts, but we will be setting them on the current machine in order to have the Kubernetes master and node configured on the same machine:**

9. **Edit the file** `/etcd/kubernetes/kubelet` **as follows:**

```
###
# Kubernetes kubelet (node) config

# The address for the info server to serve on (set to 0.0.0.0
or "" for
all interfaces)
KUBELET_ADDRESS="--address=0.0.0.0"

# You may leave this blank to use the actual hostname
KUBELET_HOSTNAME="--hostname-override=fed-node"

# location of the api-server
KUBELET_API_SERVER="--api-servers=http://fed-master:8080"

# Add your own!
#KUBELET_ARGS=""
```

10. Create a shell script to start all the Kubernetes master and node services on the same machine:

```
$ nano start-k8s.sh
for SERVICES in etcd kube-apiserver kube-controller-manager
kube-scheduler
kube-proxy kubelet docker; do
    systemctl restart $SERVICES
    systemctl enable $SERVICES
    systemctl status $SERVICES
done
```

11. Create a `node.json` file to configure it on the Kubernetes machine:

```
{
    "apiVersion": "v1",
    "kind": "Node",
    "metadata": {
        "name": "fed-node",
        "labels":{ "name": "fed-node-label"}
    },
    "spec": {
        "externalID": "fed-node"
    }
}
```

12. Create a node object using the following command:

```
$ kubectl create -f ./node.json

$ kubectl get nodes
NAME                   LABELS                      STATUS
fed-node               name=fed-node-label         Unknown
```

13. After some time, node should be ready to deploy pods:

```
kubectl get nodes
NAME                   LABELS                      STATUS
fed-node               name=fed-node-label         Ready
```

Troubleshooting the Kubernetes Fedora manual setup

If the kube-apiserver fails to start, it might be due to service account admission control and require a service account and a token before allowing pods to be scheduled. It is generated automatically by the controller. By default, the API server uses a TLS serving key, but as we are not sending over HTTPS and don't have a TLS server key, we can provide the API server the same key file in order for the API server to validate generated service-account tokens.

Use the following to generate the key and add it to the `k8s` cluster:

```
openssl genrsa -out /tmp/serviceaccount.key 2048
```

To start the API server, add the following option to the end of the `/etc/kubernetes/apiserver` file:

```
KUBE_API_ARGS="--
service_account_key_file=/tmp/serviceaccount.key"
```

`/etc/kubernetes/kube-controller-manager` add the following option to the end of the file:

```
KUBE_CONTROLLER_MANAGER_ARGS=" -
service_account_private_key_file
=/tmp/serviceaccount.key"
```

Restart the cluster using the `start_k8s.sh` shell script.

Deploying Kubernetes using Minikube

Minikube is still in development; it is a tool that makes it easy to run Kubernetes locally, optimized for the underlying OS (MAC/Linux). It runs a single-node Kubernetes cluster inside a VM. Minikube helps developers to learn Kubernetes and do day-to-day development and testing with ease.

The following setup will cover Minikube setup on Mac OS X, as very few guides are present to deploy Kubernetes on Mac:

1. Download the Minikube binary:

```
    $ curl -Lo minikube
https://storage.googleapis.com/minikube/releases/v0.12.2/minikube-darwin-am
d64
    % Total % Received % Xferd Average Speed Time Time Time Current
  Dload Upload Total Spent Left Speed
        100 79.7M 100 79.7M 0 0 1857k 0 0:00:43 0:00:43 --:--:-- 1863k
```

2. Grant execute permission to the binary:

```
$ chmod +x minikube
```

3. Move the Minikube binary to `/usr/local/bin` so that it gets added to the path and can be executed directly on the terminal:

```
    $ sudo mv minikube /usr/local/bin
```

4. After this, we'll require the `kubectl` client binary to run commands against the single-node Kubernetes cluster, for Mac OS X:

```
    $ curl -Lo kubectl
https://storage.googleapis.com/kubernetes-release/release/v1.3.0/bin/darwin
/amd64/kubectl && chmod +x kubectl && sudo mv kubectl /usr/local/bin/

https://storage.googleapis.com/kubernetes-release/release/v1.3.0/bin/darwin
/amd64/kubectl && chmod +x kubectl && sudo mv kubectl /usr/local/bin/
    % Total % Received % Xferd Average Speed Time Time Time Current
                             Dload Upload Total Spent Left Speed
        100 53.2M 100 53.2M 0 0 709k 0 0:01:16 0:01:16 --:--:-- 1723k
```

The kubectl is now configured to be used with the cluster.

5. Set up Minikube to deploy a VM locally and configure the Kubernetes cluster:

```
    $ minikube start
    Starting local Kubernetes cluster...
    Downloading Minikube ISO
    36.00 MB / 36.00 MB
[================================================]
    100.00% 0s
```

6. We can set up kubectl to use a Minikube context, and switch later if required:

```
$ kubectl config use-context minikube
switched to context "minikube".
```

7. We'll be able to list the node of the Kubernetes cluster:

```
$ kubectl get nodes
NAME        STATUS    AGE
minikube    Ready     39m
```

8. Create a `hello-minikube` pod and expose it as a service:

```
$ kubectl run hello-minikube --
    image=gcr.io/google_containers/echoserver:1.4 --port=8080
deployment "hello-minikube" created

$ kubectl expose deployment hello-minikube --type=NodePort

service "hello-minikube" exposed
```

9. We can get the `hello-minikube` pod status using the following command:

```
$  kubectl get pod
    NAME                                 READY   STATUS    RESTARTS   AGE
hello-minikube-3015430129-otr7u    1/1     running    0          36s
    vkohli-m01:~ vkohli$ curl $(minikube service hello-minikube --url)
    CLIENT VALUES:
    client_address=172.17.0.1
    command=GET
    real path=/
    query=nil
    request_version=1.1
    request_uri=http://192.168.99.100:8080/
    SERVER VALUES:
    server_version=nginx: 1.10.0 - lua: 10001
    HEADERS RECEIVED:
    accept=*/*
    host=192.168.99.100:30167
    user-agent=curl/7.43.0
```

10. We can open the Kubernetes dashboard using the following command and view details of the deployed pod:

```
$ minikube dashboard
Opening kubernetes dashboard in default browser...
```

Kubernetes UI showcasing hello-minikube pod

Deploying Kubernetes on AWS

Let's get started with Kubernetes cluster deployment on AWS, which can be done by using the configuration file already existing in the Kubernetes codebase.

1. Log in to AWS console (http://aws.amazon.com/console/)
2. Open the IAM console (https://console.aws.amazon.com/iam/home?#home)
3. Choose the IAM username, select the **Security Credentials** tab, and click the **Create Access Key** option.
4. After the keys are created, download them and keep them in a secure place. The downloaded CSV file will contain the access key ID and the secret access key, which will be used to configure the AWS CLI.

5. Install and configure the AWS command-line interface. In this example, we have installed AWS CLI on Linux using the following command:

```
$ sudo pip install awscli
```

6. In order to configure the AWS-CLI, use the following command:

```
$ aws configure
AWS Access Key ID [None]: XXXXXXXXXXXXXXXXXXXXXXXXXXXXXXXX
AWS Secret Access Key [None]: YYYYYYYYYYYYYYYYYYYYYYYYYYYYYYYY
Default region name [None]: us-east-1
Default output format [None]: text
```

7. After the configuration of the AWS CLI, we will create a profile and attach a role to it with full access to S3 and EC2.

```
$ aws iam create-instance-profile --instance-profile-name Kube
```

8. The role can be attached above the profile, which will have complete EC2 and S3 access, as shown in the following screenshot. The role can be created separately using the console or AWS CLI with the JSON file, which will define the permissions the role can have:

```
$ aws iam create-role --role-name Test-Role --assume-role-policy-
   document /root/kubernetes/Test-Role-Trust-Policy.json
```

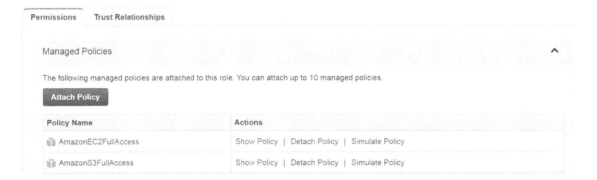

Attach policy in AWS during Kubernetes deployment

9. After the role is created, it can be attached to the policy using the following command:

```
$ aws iam add-role-to-instance-profile --role-name Test-Role --
   instance-profile-name Kube
```

10. The script uses the default profile; we can change it as follows:

    ```
    $ export AWS_DEFAULT_PROFILE=Kube
    ```

11. The Kubernetes cluster can be easily deployed using one command, as follows;

    ```
    $ export KUBERNETES_PROVIDER=aws; wget -q -O - https://get.k8s.io |
    bash
    Downloading kubernetes release v1.1.1 to
    /home/vkohli/kubernetes.tar.gz
    --2015-11-22 10:39:18--  https://storage.googleapis.com/kubernetes-
    release/release/v1.1.1/kubernetes.tar.gz
    Resolving storage.googleapis.com (storage.googleapis.com)...
    216.58.220.48, 2404:6800:4007:805::2010
    Connecting to storage.googleapis.com
    (storage.googleapis.com)|216.58.220.48|:443... connected.
    HTTP request sent, awaiting response... 200 OK
    Length: 191385739 (183M) [application/x-tar]
    Saving to: 'kubernetes.tar.gz'
    100%[======================================>] 191,385,739 1002KB/s
    in 3m 7s
    2015-11-22 10:42:25 (1002 KB/s) - 'kubernetes.tar.gz' saved
    [191385739/191385739]
    Unpacking kubernetes release v1.1.1
    Creating a kubernetes on aws...
    ... Starting cluster using provider: aws
    ... calling verify-prereqs
    ... calling kube-up
    Starting cluster using os distro: vivid
    Uploading to Amazon S3
    Creating kubernetes-staging-e458a611546dc9dc0f2a2ff2322e724a
    make_bucket: s3://kubernetes-staging-
    e458a611546dc9dc0f2a2ff2322e724a/
    +++ Staging server tars to S3 Storage: kubernetes-staging-
    e458a611546dc9dc0f2a2ff2322e724a/devel
    upload: ../../../tmp/kubernetes.6B8Fmm/s3/kubernetes-salt.tar.gz to
    s3://kubernetes-staging-
    e458a611546dc9dc0f2a2ff2322e724a/devel/kubernetes-
    salt.tar.gz
    Completed 1 of 19 part(s) with 1 file(s) remaining
    ```

12. The preceding command will call `kube-up.sh` and, in turn, the `utils.sh` using the `config-default.sh` script, which contains the basic configuration of the `k8s` cluster with four nodes, as follows:

    ```
    ZONE=${KUBE_AWS_ZONE:-us-west-2a}
    MASTER_SIZE=${MASTER_SIZE:-t2.micro}
    ```

```
MINION_SIZE=${MINION_SIZE:-t2.micro}
NUM_MINIONS=${NUM_MINIONS:-4}
AWS_S3_REGION=${AWS_S3_REGION:-us-east-1}
```

13. The instances are `t2.micro` running on Ubuntu. The process takes five to ten minutes, after which the IP address of the master and minions gets listed and can be used to access the Kubernetes cluster.

Deploying Kubernetes on vSphere

Kubernetes can be installed on vSphere with the help of `govc` (a vSphere CLI built on top of govmomi):

1. Before starting the setup, we'll have to install golang, which can be done in the following way on a Linux machine:

```
$ wget https://storage.googleapis.com/golang/go1.7.3.linux-
  amd64.tar.gz
$ tar -C /usr/local -xzf go1.7.3.linux-amd64.tar.gz
$ go
Go is a tool for managing Go source code.
Usage:
  go command [arguments]
```

2. Set the go path:

```
$ export GOPATH=/usr/local/go
$ export PATH=$PATH:$GOPATH/bin
```

3. Download the pre-built Debian VMDK, which will be used to create the Kubernetes cluster on vSphere:

```
$ curl --remote-name-all https://storage.googleapis.com/
govmomi/vmdk/2016-01-08/kube.vmdk.gz{,.md5}
    % Total    % Received % Xferd  Average Speed   Time    Time
Time
    Current
                                   Dload  Upload   Total   Spent
Left
    Speed
100   663M  100   663M    0     0  14.4M      0  0:00:45  0:00:45 --:--
:--
    17.4M
100    47  100    47    0     0     70      0 --:--:-- --:--:-- --:--
:--
```

```
0
$ md5sum -c kube.vmdk.gz.md5
kube.vmdk.gz: OK
$ gzip -d kube.vmdk.gz
```

Kubernetes setup troubleshooting

We need to set up the proper environment variables to connect remotely to the ESX server to deploy the Kubernetes cluster. The following environment variables should be set in order to progress with Kubernetes setup on vSphere:

```
export GOVC_URL='https://[USERNAME]:[PASSWORD]@[ESXI-HOSTNAME-IP]/sdk'
export GOVC_DATASTORE='[DATASTORE-NAME]'
export GOVC_DATACENTER='[DATACENTER-NAME]'
#username & password used to login to the deployed kube VM
export GOVC_RESOURCE_POOL='*/Resources'
export GOVC_GUEST_LOGIN='kube:kube'
export GOVC_INSECURE=true
```

 Use ESX and vSphere version v5.5 for this tutorial.

Upload the `kube.vmdk` to the ESX datastore. The VMDK will be stored in the `kube` directory, which will get created by the following command:

```
$ govc datastore.import kube.vmdk kube
```

Set up the Kubernetes provider as vSphere, as well the Kubernetes cluster, which will get deployed on the ESX. This will contain one Kubernetes master and four Kubernetes minion derived from the expanded `kube.vmdk` uploaded in the datastore:

```
$ cd kubernetes
$ KUBERNETES_PROVIDER=vsphere cluster/kube-up.sh
```

This will display a list of IP addresses for the four VMs. If you are currently developing Kubernetes, you can use this cluster-deployment mechanism to test out the new code in the following way:

```
$ cd kubernetes
$ make release
$ KUBERNETES_PROVIDER=vsphere cluster/kube-up.sh
```

The cluster can be brought down using the following command:

```
$ cluster/kube-down.sh
```

What is a Virtual Machine?

A virtual machine is a software computer that, like a physical computer, runs an operating system and applications. An operating system installed on a virtual machine is called a guest operating system.

Virtual Machines

Kubernetes master/node deployed on vSphere

Kubernetes pod deployment

Now, in the following example, we will be deploying two NGINX replication pods (rc-pod) and exposing them via a service. To understand Kubernetes networking, please refer to the following diagram for more details. Here, an application can be exposed via a virtual IP address, and the request to be proxied, to which replica of pod (load balancer), is taken care of by the service:

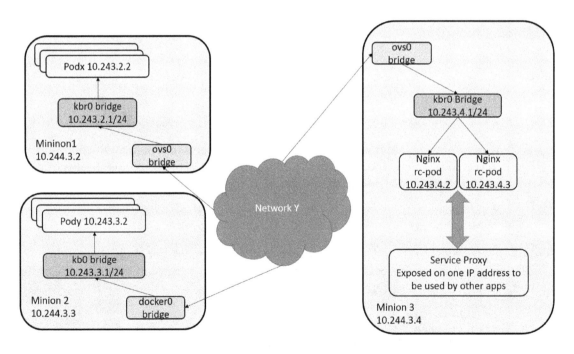

Kubernetes networking with OVS bridge

1. In the Kubernetes master, create a new folder:

   ```
   $ mkdir nginx_kube_example
   $ cd nginx_kube_example
   ```

2. Create the YAML file in the editor of your choice, which will be used to deploy the NGINX pod:

   ```
   $ vi nginx_pod.yaml
   apiVersion: v1
   kind: ReplicationController
   metadata:
     name: nginx
   spec:
     replicas: 2
     selector:
       app: nginx
     template:
       metadata:
         name: nginx
         labels:
           app: nginx
     spec:
       containers:
       - name: nginx
         image: nginx
         ports:
         - containerPort: 80
   ```

3. Create the NGINX pod using `kubectl`:

   ```
   $ kubectl create -f nginx_pod.yaml
   ```

4. In the preceding pod creation process, we have created two replicas of the NGINX pod, and its details can be listed as shown here:

   ```
   $ kubectl get pods
   NAME            READY       REASON      RESTARTS    AGE
   nginx-karne     1/1         Running     0           14s
   nginx-mo5ug     1/1         Running     0           14s
   $ kubectl get rc
   CONTROLLER      CONTAINER(S)    IMAGE(S)    SELECTOR    REPLICAS
   nginx           nginx           nginx       app=nginx   2
   ```

5. The container on the deployed minion can be listed as follows:

```
$ docker ps
CONTAINER ID          IMAGE                                    COMMAND
CREATED               STATUS              PORTS                NAMES
1d3f9cedff1d          nginx:latest                             "nginx
-g
'daemon of    41 seconds ago        Up 40 seconds
k8s_nginx.6171169d_nginx-karne_default_5d5bc813-3166-11e5-8256-
ecf4bb2bbd90_886ddf56
0b2b03b05a8d          nginx:latest                             "nginx
-g
'daemon of    41 seconds ago        Up 40 seconds
```

6. Deploy the NGINX service using a YAML file in order to expose the NGINX pod on host port 82:

```
$ vi nginx_service.yaml
apiVersion: v1
kind: Service
metadata:
  labels:
    name: nginxservice
  name: nginxservice
spec:
  ports:
    # The port that this service should serve on.
    - port: 82
    # Label keys and values that must match in order to receive
traffic for
  this service.
    selector:
      app: nginx
    type: LoadBalancer
```

7. Create the NGINX service using `kubectl`:

```
$kubectl create -f nginx_service.yaml
services/nginxservice
```

8. The NGINX service can be listed as follows:

```
$ kubectl get services
NAME                  LABELS                                   SELECTOR
IP(S)
                      PORT(S)
kubernetes            component=apiserver,provider=kubernetes  <none>
```

```
192.168.3.1      443/TCP
nginxservice     name=nginxservice                        app=nginx
192.168.3.43     82/TCP
```

9. Now the NGINX server test page via service can be accessed on the following
URL: `http://192.168.3.43:82`

Deploying Kubernetes in a production environment

In this section, we'll be covering some of the important points and concepts that can be used
to deploy Kubernetes in production.

- **Exposing Kubernetes services**: Once we deploy the Kubernetes pods, we expose
  them using services. The Kubernetes service is an abstraction, which defines a set
  of pods and a policy to expose them as a microservice. The service gets its own IP
  address, but the catch is that this address only exists within the Kubernetes
  cluster, which means the service is not exposed to the Internet.
  It's possible to expose the service directly on the host machine port, but once we
  expose the service on the host machine, we get into port conflicts. It also voids
  Kubernetes benefits and makes it harder to scale the deployed service:

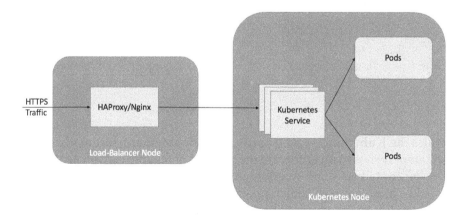

Kubernetes service exposed through external load balancer

One solution is to add an external load balancer such as HAProxy or NGINX. This is configured with a backend for each Kubernetes service and proxies traffic to individual pods. Similar to AWS deployment, a Kubernetes cluster can be deployed inside a VPN and an AWS external load balancer can be used to expose each Kubernetes service:

- **Support upgrade scenarios in Kubernetes**: In the case of an upgrade scenario, we need to have zero downtime. Kubernetes' external load balancer helps to achieve this functionality in cases of service deployment through Kubernetes. We can start a replica cluster running the new version of the service, and the older cluster version will serve the live requests. As and when the new service is ready, the load balancer can be configured to switch load to the new version. By using this approach, we can support a zero-runtime upgrade scenario for enterprise products:

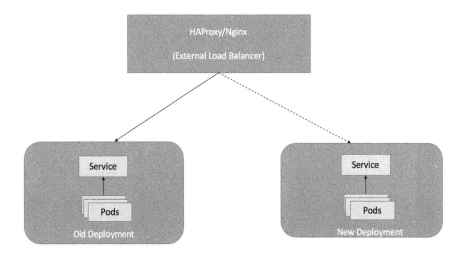

Upgrade scenarios supported in Kubernetes deployment

- **Make the Kubernetes-based application deployment automatic**: With the help of a deployer, we can automate the process of testing, as well as deploying the Docker containers in production. In order to do so, we need to have a build pipeline and deployer, which pushes the Docker image to a registry such as Docker Hub after successful build. Then, the deployer will take care of deploying the test environment and invoke the test scripts. After successful testing, the deployer can also take care of deploying the service in the Kubernetes production environment:

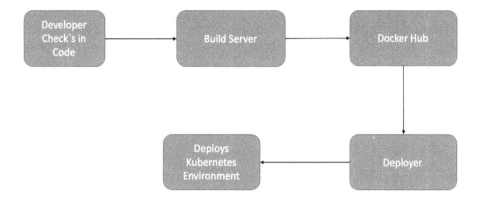

Kubernetes application deployment pipeline

- **Know the resource constraints**: Know the resource constraints while starting Kubernetes cluster, configure the resource requests and CPU/memory limits on each pod. Most containers crash in the production environment due to lack of resources or insufficient memory. The containers should be well tested, and the appropriate resource should be allotted to the pod in the production environment for successful deployment of the microservice.

- **Monitor the Kubernetes cluster**: The Kubernetes cluster should be continuously monitored with the help of logging. Logging tools such as Graylog, Logcheck, or Logwatch should be used with Apache Kafka, a messaging system to collect logs from the containers and direct them to the logging tools. With the help of Kafka, it is easy to index the logs, as well as handle huge streams. Kubernetes replica works flawlessly. If any pod crashes, the Kubernetes service restarts them and keeps the number of replicas always up and running as per the configuration. One aspect that users like to know about is the real reason behind the failure. Kubernetes metrics and application metrics can be published to a time-series store such as InfluxDB, which can be used to track application errors and measure load, throughput, and other stats to perform post-analysis of the failure.

- **Persistent storage in Kubernetes**: Kubernetes has the concept of volumes to work with persistent data. We want persistence storage in a production deployment of Kubernetes because containers lose their data as they restart. A volume is backed by a variety of implementations, such as host machines, NFS, or using the cloud-provider volume service. Kubernetes also provides two APIs to handle persistent storage:

- **Persistent volume (PV)**: This is a resource, provisioned in a cluster, which behaves as though a node is a cluster resource. Pods request the resource (CPU and memory) as required from the persistent volumes. It is usually provisioned by an administrator.
- **Persistent volume claim (PVC)**: A PVC consumes PV resources. It is a request for storage by the user, similar to a pod. A pod can request levels of resources (CPU and memory) as required.

Debugging Kubernetes issues

In this section, we'll be discussing some of the Kubernetes troubleshooting concerns:

1. The first step to debug the Kubernetes cluster is to list the number of nodes, using the following command:

   ```
   $ kubetl get nodes
   ```

   Also, verify that all nodes are in the ready state.

2. Look at the logs in order to figure out issues in the deployed Kubernetes cluster

   ```
   master:
       var/log/kube-apiserver.log - API Server, responsible for serving
   the API
       /var/log/kube-scheduler.log - Scheduler, responsible for making
   scheduling
       decisions
       /var/log/kube-controller-manager.log - Controller that manages
   replication
       controllers
       Worker nodes:
       /var/log/kubelet.log - Kubelet, responsible for running containers
   on the
       node
       /var/log/kube-proxy.log - Kube Proxy, responsible for service load
       balancing
   ```

3. If the pod stays in the pending state, use the following command:

```
$ cluster/kubectl.sh describe pod podname
```

This will list events and might describe the last thing that happened to the pod.

4. To see all the cluster events, use the following command:

```
$ cluster/kubectl.sh get events
```

If the `kubectl` command line is unable to reach the `apiserver` process, ensure `Kubernetes_master` or `Kube_Master_IP` is set. Ensure the `apiserver` process is running in the master and check its logs:

- If you are able to create the replication controller but not see the pods: If the replication controller didn't create the pods, check if the controller is running and look at the logs.
- If `kubectl` hangs forever or a pod is in the waiting state:
    - Check if the hosts are being assigned to the pod, if not then currently they are being scheduled for some task.
    - Check if the kubelet is pointing at the right place in `etcd` for pods and the `apiserver` is using the same name or IP of the minion.
    - Check if the Docker daemon is running if some issue occurs. Also, check the Docker logs and make sure the firewall is not blocking the image from being fetched from Docker Hub.
- The `apiserver` process reports:
- Error synchronizing container: `Get http://:10250/podInfo?podID=foo: dial tcp :10250:` **connection refused**:
    - This means the pod has not yet been scheduled
    - Check the scheduler logs to see if it is running properly
    - Cannot connect to the container
    - Try to Telnet to the minion at the service port or the pod's IP address

- Check if the container is created in Docker using the following command:

```
$ sudo docker ps -a
```

- If you don't see the container, the issue will be with the pod configuration, image, Docker, or the kubelet. If you see the container getting created every 10 seconds, then the issues are with the container creation, or the container's process is failing.
- X.509 certificate has expired or is not yet valid.

Check if the current time matches on the client and server. Use `ntpdate` for one-time clock synchronization.

Summary

In this chapter, we learned about managing Docker containers with help of Kubernetes. Kubernetes have a different perspective among Docker orchestration tools, where each pod will get a unique IP address and communication between pods can occur with the help of services. We have covered many deployment scenarios, as well as troubleshooting issues when deploying Kubernetes on a Bare Metal machine, AWS, vSphere, or using Minikube. We also looked at deploying Kubernetes pods effectively and debugging Kubernetes issues. The final section helps with deploying Kubernetes in a production environment with load balancers, Kubernetes services, monitoring tools, and persistent storage. In the next chapter, we will cover Docker volumes and how to use them efficiently in a production environment.

9
Hooking Volume Baggage

This chapter introduces data volumes and storage driver concepts, which are widely used in Docker to manage persistent or shared data. We'll be also taking a deep dive into various storage drivers supported by Docker, and the basic commands associated with them for management. The three main use cases for Docker data volumes are as follows:

- To keep data persistent after a container is deleted
- To share data between the host and the Docker container
- To share data across Docker containers

In order to understand a Docker volume, we need to understand how the Docker filesystem works. Docker images are stored as a series of read-only layers. When the container is started, the read-only image adds a read-write layer on top. If the current file needs to be modified, it is copied from the read-only layer to the read-write layer, where changes are applied. The version of the file in the read-write layer hides the underlying file but doesn't destroy it. Thus, when a Docker container is deleted, relaunching the image will start a fresh container with a fresh read-write layer and all the changes are lost. The combination of read-write layers on top of the read-only layer is termed the **Union File System** (**UFS**). In order to persist the data and be able to share it with the host and other containers, Docker has come up with the concept of volumes. Basically, volumes are directories that exist outside the UFS and behave as normal directories or files on the host filesystem.

Some important features of Docker volumes are as follows:

- Volumes can be initialized when the container is created
- Data volumes can be reused and shared among other data containers
- Data volumes persist the data even if a container is deleted
- Changes to the data volume are made directly, bypassing the UFS

In this chapter, we will cover the following:

- Data-only containers
- Hosting a mapped volume backed up by shared storage
- Docker storage driver performance

Avoiding troubleshooting by understanding Docker volumes

In this section, we'll be looking at four ways to deal with data and Docker containers, which will help us to understand and achieve the preceding use cases mentioned with Docker volumes.

Default case storing data inside the Docker container

In this case, data is only visible inside the Docker containers and is not from the host system. The data is lost if the container is shut down or the Docker host dies. This case mostly works with services that are packaged in Docker containers and are not dependent on persistent data when they return:

```
$ docker run -it ubuntu:14.04
root@358b511effb0:/# cd /tmp/
root@358b511effb0:/tmp# cat > hello.txt
hii
root@358b511effb0:/tmp# ls
hello.txt
```

As seen in the preceding example, the hello.txt file only exists inside the container and will not be persisted once the container dies:

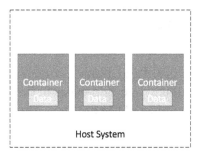

Data stored inside Docker Container

Data-only container

Data can be stored outside the Docker UFS in a data-only container. The data will be visible inside the data-only container mount namespace. As the data is persisted outside the container, it remains even after the container is deleted. If any other container wants to connect to this data-only container, simply use the `--volumes-from` option to grab the container and apply it to the current container. Let's try out data volume container:

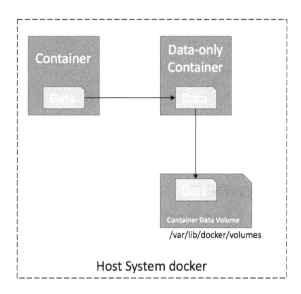

Using a data-only container

Creating a data-only container

```
$ docker create -v /tmp --name ubuntuvolume Ubuntu:14.04
```

In the preceding command, we created an Ubuntu container and attached /tmp. It is a data-only container based on the Ubuntu image, and exists in the /tmp directory. If the new Ubuntu container needs to write some data to the /tmp directory of our data-only container, this can be achieved with help of --volumes-from option. Now, anything we write to the /tmp directory of the new container will be saved in the /tmp volume of the Ubuntu data container:

```
$ docker create -v /tmp --name ubuntuvolume ubuntu:14.04
d694752455f7351e95d1563ed921257654a1867c467a2813ae25e7d99c067234
```

Use a data-volume container in container-1:

```
$ docker run -t -i --volumes-from ubuntuvolume ubuntu:14.04 /bin/bash
root@127eba0504cd:/# echo "testing data container" > /tmp/hello
root@127eba0504cd:/# exit
exit
```

Use a data-volume container in container-2 to get the data shared by container-1:

```
$ docker run -t -i --volumes-from ubuntuvolume ubuntu:14.04 /bin/bash
root@5dd8152155de:/# cd tmp/
root@5dd8152155de:/tmp# ls
hello
root@5dd8152155de:/tmp# cat hello
testing data container
```

As we can see, container-2 gets the data written by container-1 in the /tmp space. These examples demonstrate the basic usage of data-only containers.

Sharing data between the host and the Docker container

This is a common use case where it is necessary to share files between the host and the Docker container. In this scenario, we don't need to create a data-only container; we can simply run a container of any Docker image and simply override one of its directories with content from the host system directory.

Let's consider an example where we want to access the logs of Docker NGINX from the host system. Currently, they are not available outside the host, but this can be achieved simply by mapping the /var/log/nginx from inside the container to a directory on the host system. In this scenario, we will run a copy of the NGINX image with a shared volume from the host system, as shown here:

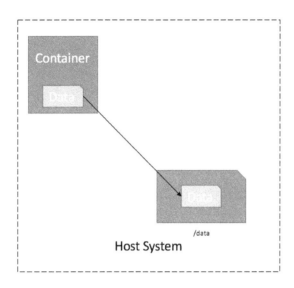

Sharing data between the host and Docker container

Create a serverlogs directory in the host system:

```
$ mkdir /home/serverlogs
```

Run the NGINX container and map /home/serverlogs to the /var/log/nginx directory inside the Docker container:

```
$ docker run -d -v /home/serverlogs:/var/log/nginx -p 5000:80 nginx
Unable to find image 'nginx:latest' locally
latest: Pulling from library/nginx
5040bd298390: Pull complete
. . .
```

Access `http://localhost:5000` from the host system, post this, logs will be generated, and they can be accessed on the host system in `/home/serverlogs` directory, which is mapped to `/var/log/nginx` inside the Docker container, as shown here:

```
$ cd serverlogs/
$ ls
access.log  error.log
$ cat access.log
172.17.42.1 - - [20/Jan/2017:14:57:41 +0000] "GET / HTTP/1.1" 200 612 "-"
"Mozilla/5.0 (X11; Ubuntu; Linux x86_64; rv:50.0) Gecko/20100101
Firefox/50.0" "-"
```

Host mapped volume backed up by shared storage

Docker volume plugins allow us to mount a shared storage backend. The main advantage of this is that the user will never suffer data loss in the case of host failure, as it is backed by shared storage. In the preceding approaches, if we migrate the container, the volumes doesn't get migrated. It can be achieved with the help of external Docker volume plugins such **Flocker** and **Convy**, which make the volume portable and help to migrate the containers across hosts with volumes easily, as well as protecting the data, as it is not dependent on the host file system.

Flocker

Flocker is widely used to run containerized stateful services and applications that require persistent storage. Docker provides a very basic view of volume management, but Flocker enhances it by providing durability, failover, and high availability of the volumes. Flocker can be deployed manually with Docker Swarm and compose, or can be set up easily on AWS with the help of the CloudFormation template if the backed up storage has to be used in production set ups.

Flocker can be deployed easily on AWS with the help of the following steps:

1. Log in to your AWS account and create a key pair in Amazon EC2.
2. Select **CloudFormation** from the home page of AWS.
3. The Flocker cloud formation stack can be launched with the help of the template in the AWS S3 storage using the following link:
   ```
   https://s3.amazonaws.com/installer.downloads.clusterhq.com/floc
   ker-cluster.cloudformation.json
   ```
4. Select create stack; then select the second option and specify the Amazon S3 template URL:

Create stack

Select Template
Specify Details
Options
Review

Select Template

Select the template that describes the stack that you want to create. A stack is a group of related resources that you manage as a single unit.

Design a template   Use AWS CloudFormation Designer to create or modify an existing template. Learn more.

Design template

Choose a template   A template is a JSON/YAML-formatted text file that describes your stack's resources and their properties. Learn more.

Select a sample template

Upload a template to Amazon S3
Choose File   No file chosen

Specify an Amazon S3 template URL

https://s3.amazonaws.com/installer.downloads.clusterhq.com/flocker-   View/Edit template in Designer

Cancel   Next

5. On the next screen, specify the **Stack name**, **AmazonAccessKeyID**, and **AmazonSecretAccessKey** for the account:

6. Provide the key-value pairs to tag this Flocker stack, and provide the **IAM Role** for this stack if required:

7. Review the details and launch the Flocker cloud formation stack:

8. Once, the stack deployment is completed from the outputs tab, get the IP address of the client node and control node. SSH into the client node using the key-value pair generated during the start of the Flocker stack deployment.

Set the following parameters:

```
$ export FLOCKER_CERTS_PATH=/etc/flocker
$ export FLOCKER_USER=user1
$ export FLOCKER_CONTROL_SERVICE=<ControlNodeIP> # not ClientNodeIP!
$ export DOCKER_TLS_VERIFY=1
$ export DOCKER_HOST=tcp://<ControlNodeIP>:2376
$ flockerctl status # should list two servers (nodes) running
$ flockerctl ls # should display no datasets yet
$ docker info |grep Nodes # should output "Nodes: 2"
```

If the Flocker `status` and `ls` commands ran successfully, this means the Docker Swarm and Flocker have been successfully set up on the AWS.

The Flocker volume can be easily set up and allows you to create a container that will persist beyond the lifecycle of the container or container host:

```
$ docker run --volume-driver flocker -v flocker-volume:/cont-dir --
name=testing-container
```

An external storage block will be created and mounted, to our host and the container directory will be bounded to it. If the container is deleted or the host crashes, the data remains secured. The alternate container can be brought up in the second host using the same command, and we will be able to access our shared storage. The preceding tutorial was to set up Flocker on the AWS for a production use case, but we can also test Flocker locally with the help of Docker Swarm setup. Let us consider a use case where you have two Docker Swarm nodes and a Flocker client node.

In the Flocker client node

Create a `docker-compose.yml` file and define the containers `redis` and `clusterhq/flask`. Provide the respective configuration Docker image, names, ports, and data volumes:

```
$ nano docker-compose.yml

web:
  image: clusterhq/flask
  links:
   - "redis:redis"
  ports:
   - "80:80"
redis:
  image: redis:latest
  ports:
   - "6379:6379"
  volumes: ["/data"]
```

Create a file named `flocker-deploy.yml`, where we will define both containers that will be deployed on the same nodes–`node-1`; leave `node-2` blank as of now of the Swarm cluster:

```
$ nano flocker-deploy.yml
"version": 1
"nodes":
```

```
"node-1": ["web", "redis"]
"node-2": []
```

Deploy the containers using the preceding `.yml` files; we simply need to run the following command to do so:

```
$ flocker-deploy control-service flocker-deploy.yml docker-compose.yml
```

The cluster configuration has been updated. It may take a short while for the changes to take effect, in particular if Docker images need to be pulled.

Both containers can be observed running in `node-1`. Once the setup has been done, we can access the application on `http://node-1`. It will show the visit count of this webpage:

```
"Hello... I have been seen 8 times"
```

Recreate the deployment file in order to move the container to `node-2`:

```
$ nano flocker-deploy-alt.yml
"version": 1.
"nodes":
  "node-1": ["web"]
  "node-2": ["redis"]
```

Now, we'll be migrating the container from `node-1` to `node-2`, and we'll see that Flocker will auto handle the volume management. It will plug the existing volume to the Redis container when it comes up in `node-2`:

```
$ flocker-deploy control-service flocker-deploy-alt.yml docker-compose.yml
```

The cluster configuration has been updated. It may take a short while for the changes to take effect, in particular if Docker images need to be pulled.

We can SSH into `node-2` and list the running Redis container. Try to access the application on `http://node2`; we'll be able to see that the count is still persisted as it were in `node-1` and gets incremented by 1 as the application is accessed from `node-2`:

```
"Hello... I have been seen 9 times"
```

This example demonstrates how easily we can migrate the container with its data volume in a Flocker cluster from one node to another.

Convoy Docker volume plugin

Convoy is the other Docker volume plugin that is widely used to provide storage backend. It is written in Go and the main advantage is that it can be deployed in standalone mode. Convoy will run as a Docker volume extension, and will behave like an intermediate container. The initial implementation of Convoy utilizes Linux devices and provides the following four Docker storage function for volumes:

- Thin provisioned volumes
- Restore volumes across hosts
- Take snapshots of volumes
- Back up the volumes to external object stores such as **Amazon EBS**, **Virtual File System** (**VFS**), and **Network File System** (**NFS**):

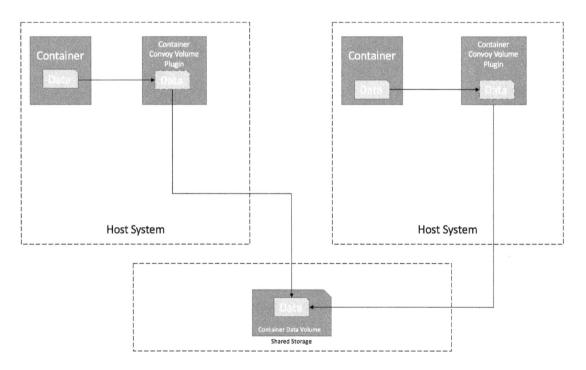

Using Convoy volume plugin

In the following example, we'll be running a local Convoy device mapper driver and showcasing the use of the Convoy volume plugin in between two containers for sharing the data:

1. Verify the Docker version is above 1.8.
2. Install the Convoy plugin by locally downloading the plugin tar file and extracting it:

```
$ wget https://github.com/rancher/convoy/releases/download
/v0.5.0/convoy.tar.gz
$ tar xvf convoy.tar.gz
convoy/
convoy/convoy-pdata_tools
convoy/convoy
convoy/SHA1SUMS
$ sudo cp convoy/convoy convoy/convoy-pdata_tools /usr/local/bin/
$ sudo mkdir -p /etc/docker/plugins/
$ sudo bash -c 'echo "unix:///var/run/convoy/convoy.sock" >
/etc/docker/plugins/convoy.spec'
```

3. We can go ahead and use the file backed loop device which acts as a pseudo device and makes file accessible as a block device in order to demo the Convoy device mapper driver:

```
$ truncate -s 100G data.vol
$ truncate -s 1G metadata.vol
$ sudo losetup /dev/loop5 data.vol
$ sudo losetup /dev/loop6 metadata.vol
```

4. Once the data and metadata device is setup start Convoy plugin daemon:

```
sudo convoy daemon --drivers devicemapper --driver-opts
dm.datadev=/dev/loop5 --driver-opts dm.metadatadev=/dev/loop6
```

5. In the preceding Terminal, the Convoy daemon will start running; open the next Terminal instance and create a `busybox` Docker container, which uses the Convoy volume `test_volume` mounted at `/sample` directory inside the container:

```
$ sudo docker run -it -v test_volume:/sample --volume-driver=convoy
busybox
Unable to find image 'busybox:latest' locally
latest: Pulling from library/busybox
4b0bc1c4050b: Pull complete
Digest: sha256:817a12c32a39bbe394944ba49de563e085f1d3c5266eb8
e9723256bc4448680e
```

```
Status: Downloaded newer image for busybox:latest
```

6. Create a sample file in the mounted directory:

    ```
    / # cd sample/
    / # cat > test
    testing
    /sample # exit
    ```

7. Start a different container by using the volume driver as Convoy and mount the same Convoy volume:

    ```
    $ sudo docker run -it -v test_volume:/sample --volume-driver=convoy --
    name=new-container busybox
    ```

8. As we do `ls`, we'll be able to see the file created in the previous container:

    ```
    / # cd sample/
    /sample # ls
    lost+found   test
    /sample # exit
    ```

Thus, the preceding example shows how Convoy can allow the sharing of volumes between containers residing in the same, or a different, host.

Basically, the volume driver should be used for persistent data such as WordPress MySQL DB:

```
$ docker run --name wordpressdb --volume-driver=convoy -v
test_volume:/var/lib/mysql -e MYSQL_ROOT_PASSWORD=password -e
MYSQL_DATABASE=wordpress -d mysql:5.7
1e7908c60ceb3b286c8fe6a183765c1b81d8132ddda24a6ba8f182f55afa2167

$ docker run -e WORDPRESS_DB_PASSWORD=password -d --name wordpress --link
wordpressdb:mysql   wordpress
0ef9a9bdad448a6068f33a8d88391b6f30688ec4d3341201b1ddc9c2e641f263
```

In the preceding example, we have started the MySQL DB using the Convoy volume driver in order to provide persistence in case the host fails. We then linked the MySQL database in the WordPress Docker container.

Docker storage driver performance

In this section, we'll be looking into the performance aspect and comparison of file systems supported by Docker. Pluggable storage driver architecture and the flexibility to plug in a volume is the best approach for containerized environments and production use cases. Docker supports the aufs, btrfs, devicemapper, vfs, zfs, and overlayfs filesystems.

UFS basics

As discussed previously, Docker uses UFS in order to have a read-only, layered approach.

Docker uses UFS to combine several such layers into a single image. This section will take a deep dive into the basics of UFS and storage drivers supported by Docker.

UFS recursively merges several directories into a single virtual view. The fundamental desire of UFS is to have a read-only file system and some writable overlay on it. This gives the illusion that the file system has read-write access, even though it is read-only. UFS uses copy-on-write to support this feature. Also, UFS operates on directories instead of drives.

The underlying filesystem does not matter. UFS can combine directories from different underlying file systems. Combining different underlying filesystems is possible because UFS intercepts the operations bound to those file systems. The following diagram shows that the UFS lies between the user applications and filesystems. Examples of UFS are Union FS, Another Union FS (AUFS), and so on:

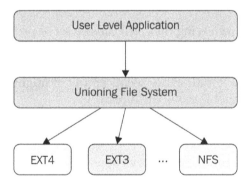

UFS and underlying file systems for branches

UFS – terminology

Branches in UFS are filesystems that are merged. Branches can have different access permissions, such as read-only, read-write, and so on. UFSs are stackable filesystems. The branches can also be assigned preferences, which determine the order in which operations will be performed on the filesystems. If a directory with the same file name exists in multiple branches, the contents of the directory appear to be merged in the UFS, but the operations on the files in those directories are redirected to respective filesystems.

UFS allows us to create a writable layer over a read-only file system and create new files/directories. It also allows the updating of existing files. The existing files are updated by copying the file to the writable layer and then making the changes. A file in a read-only file system is kept as it is, but the virtual view created by UFS will show an updated file. This phenomenon of copying a file to a writable layer to update it is called copy-up.

With copy-up in place, removing files becomes complex. When trying to delete a file, we have to delete all the copies from bottom to top. This can result in errors on read-only layers, which cannot remove the file. In such situations, the file is removed from writable layers, but still exists in the read-only layers below.

UFS – issues

The most obvious problem with UFS is support for underlying filesystems. Since UFS wraps the necessary branches and their filesystems, the filesystem support has to be added in the UFS source code. The underlying filesystems do not change, but UFS has to add support for each one of them.

The whiteouts created after removing files also cause a lot of problems. First and foremost is that they pollute the filesystem namespace. This can be reduced by adding whiteouts in a single sub-directory, but that needs special handling. Also, because of whiteouts, `rmdir` performance degrades. Even if a directory seems empty, it might contain a lot of whiteouts, because of which `rmdir` cannot remove the directory.

Copy-up is an excellent feature in UFS, but it also has drawbacks. It reduces the performance for the first update, as it has to copy the complete file and directory hierarchy to a writable layer. Also, the time of directory copies needs to be decided. There are two choices: copy the whole directory hierarchy while updating, or do it when the directory is opened. Both techniques have their trade-offs.

AuFS

AuFS is another UFS. AuFS is forked from the UFS file system. This caught the eye of developers, and is now way ahead of UFS. In fact, UFS is now following some of the design decisions taken while developing AuFS. Like any UFS, AuFS makes existing filesystem and overlays a unified view on it.

AuFS supports all the UFS features mentioned in the previous sections. You need to install the `aufs-tools` package on Ubuntu to use AuFS commands. More information about AuFS and its commands can be found on the AuFS man page.

Device Mapper

Device Mapper is a Linux kernel component; it provides a mechanism for mapping physical block devices onto virtual block devices. These mapped devices can be used as logical volumes. Device Mapper provides a generic way to create such mappings.

Device Mapper maintains a table, which defines device mappings. The table specifies how to map each range of logical sectors of the device. The table contains lines for the following parameters:

- `start`
- `length`
- `mapping`
- `mapping_parameters`

The `start` value for the first line is always zero. For other lines, start plus the length of the previous line should be equal to the `start` value of the current line. Device Mapper sizes are always specified in 512 byte sectors. There are different types of mapping targets, such as linear, striped, mirror, snapshot, snapshot-origin, and so on.

How device-mapper is used by Docker

Docker uses the thin provisioning and snapshots features of Device Mapper. These features allow many virtual devices to be stored on the same data volume. Two separate devices are used for data and metadata. The data device is utilized for the pool itself and the metadata device contains information about volumes, snapshots, blocks in the storage pool, and mapping between the blocks of each snapshot. So, Docker creates a single large block device on which a thin pool is created. It then creates a base block device. Every image and container is formed from the snapshot of this base device.

BTRFS

BTRFS is a Linux filesystem that has the potential to replace the current default Linux filesystem, EXT3/EXT4. BTRFS (also known as **butter FS**) is basically a copy-on-write filesystem. **Copy-on-Write (CoW)** means it never updates the data. Instead, it creates a new copy of that part of the data which is stored somewhere else on the disk, keeping the old part as it is. Anyone with decent filesystem knowledge will understand that CoW requires more space because it stores the old copies of data as well. Also, it has the problem of fragmentation. So, how can a CoW filesystem be used as a default Linux filesystem? Wouldn't that reduce the performance? No need to mention the storage space problem. Let's dive into BTRFS to understand why it has become so popular.

The primary design goal of BTRFS was to develop a generic filesystem that can perform well with any use case and workload. Most filesystems perform well for a specific filesystem benchmark, and performance is not that great for other scenarios. Apart from this, BTRFS also supports snapshots, cloning, and RAID (Level 0, 1, 5, 6, 10). This is more than anyone has previously bargained for from a filesystem. One can understand the design complexity, because Linux filesystems are deployed on all kinds of devices, from computers and smart phones to small embedded devices. The BTRFS layout is represented with B-trees, more like a forest of B-trees. These are copy-on-write-friendly B-trees. As CoW filesystems require a little more disk space, in general, BTRFS has a very sophisticated mechanism for space reclamation. It has a garbage collector, which makes use of reference counting to reclaim unused disk space. For data integrity, BTRFS uses check sums.

The storage driver can be selected by passing the `--storage-driver` option to the `dockerd` command line, or setting the `DOCKER_OPTS` option in the `/etc/default/docker` file:

```
$ dockerd --storage-driver=devicemapper &
```

We have considered the preceding three widely used filesystems with Docker in order to do performance analysis for the following Docker commands using micro benchmark tools; `fio` is the tool used to analyze the details of the filesystem, such as random write:

- `commit`: This is used to create a Docker image out of a running container:

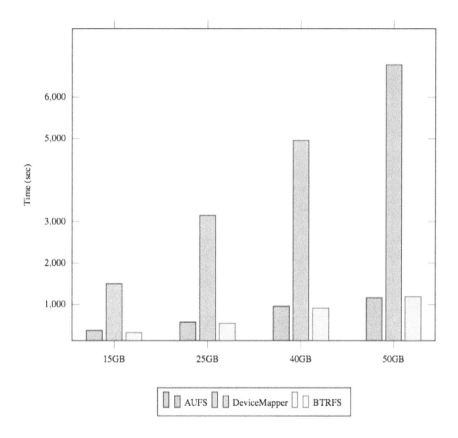

Chart depicting the time required to commit a large-size container containing a single large file

- `build`: This is used to build an image from using a Dockerfile which contains a set of steps to be followed to create an image from scratch containing a single large file:

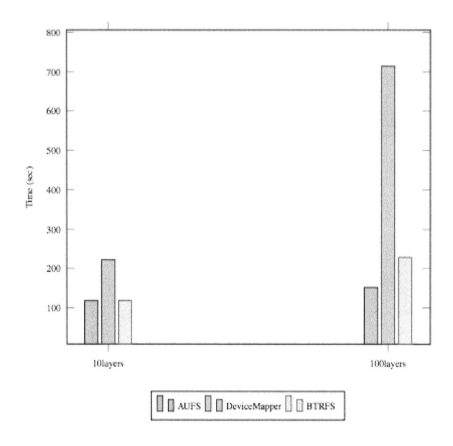

Chart depicting the time required to build the container on different file systems

- `rm`: This is used to remove a stopped container:

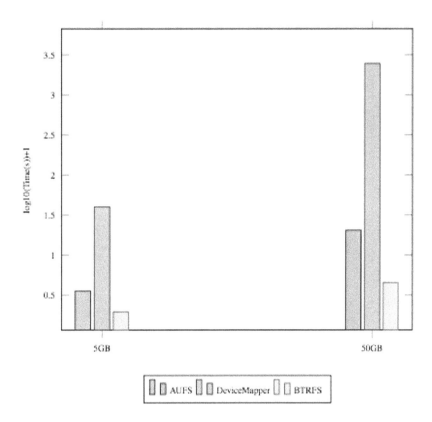

Chart depicting the time required to remove the container holding many thousands of files using the rm command

- `rmi`: This is used to remove an image:

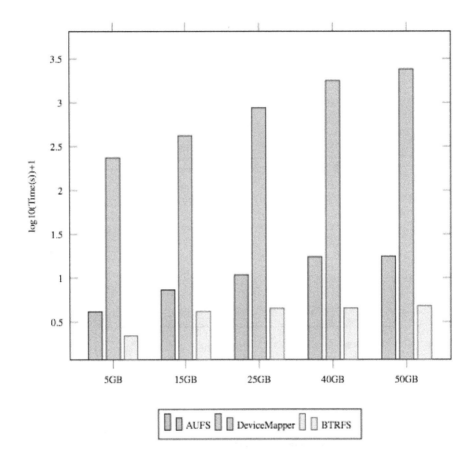

Chart depicting the time required to remove a large size container containing a single large file using the rmi command

From the preceding tests, we can clearly see that AuFS and BTRFS perform extremely well for Docker commands, but BTRFS containers performing many small writes leads to poor use of the BTRFS chunk. This can ultimately lead to out-of-space conditions on the Docker host and stop working. Using the BTRFS storage driver closely monitors the free space on the BTRFS filesystem. Also, due to the BTRFS journaling technique, the sequential writes are affected and can halve performance.

Device Mapper performs badly, as each time the container updates existing data, the storage driver performs a CoW operation. The copy is from the image snapshot to the container's snapshot and can have a noticeable impact on container performance.

AuFS looks like a good choice for PaaS and other similar use-cases where container density plays an important role. AuFS efficiently shares images between running, enabling a fast container start time and minimal use of disk space. It also uses system page cache very efficiently. OverlayFS is a modern filesystem similar to AuFS, but with a simpler design and potentially faster. But currently, OverlayFS is not mature enough to be used in a production environment. It may be a successor to AuFS in the near future. No single driver is well suited for every use case. Users should either select the storage driver as per the use case and considering the stability required for the application, or go ahead with the default driver installed by the distribution's Docker package. If the host system is RHEL or a variation, Device Mapper is the default storage driver. For Ubuntu, AuFS is the default driver.

Summary

In this chapter, we took a deep dive into data volumes and storage driver concepts related to Docker. We discussed troubleshooting the data volumes with the help of the four approaches, as well as their pros and cons. The first case of storing data inside the Docker container is the most basic case, but doesn't provide the flexibility to manage and handle data in a production environment. The second and third cases are about storing the data using data-only containers or directly on the host. These cases help to provide reliability, but still depend on the availability of host. The fourth case, which is about using a third-party volume plugin such as Flocker or Convoy, solves all of the preceding issues by storing the data in a separate block, and provides the reliability with data, even if the container is transferred from one host to another or if the container dies. In the final section we discussed Docker storage drivers and the plugin architecture provided by Docker to use required filesystems such as AuFS, BTRFS, Device Mapper, vfs, zfs and OverlayFS. We looked in depth at AuFS, BTRFS, and Device Mapper, which are widely used filesystems. From the various tests we conducted using the basic Docker commands, AuFS and BTRFS provide a better performance than Device Mapper. Users should select a Docker storage driver as per their application use case and Docker daemon host system.

In the next chapter, we'll discuss Docker deployment in a public cloud, AWS and Azure, and troubleshooting issues.

10
Docker Deployment in a Public Cloud - AWS and Azure

In this chapter, we'll be doing Docker deployment on public clouds AWS and Azure. AWS rolled out the **Elastic Compute Cloud** (**EC2**) container service towards the end of 2014. When it was launched, the company emphasised the management tasks with container cluster management with high-level APIs calls based on Amazon services release in the past. AWS has recently released Docker for AWS Beta, which allows users to quickly set up and configure a Docker 1.13 swarm mode on AWS as well as on Azure. With the help of this new service, we get the following features:

- It ensures teams to can seamlessly move apps from the developer's laptop to a Dockerised staging and production environment
- It helps to deeply integrate with underlying AWS and Azure infrastructure, takes advantage of the host environment, and exposes familiar interfaces to administrators using the public cloud
- It deploys the platform and migrates easily across various platforms where Dockerised apps can be moved simply and efficiently
- It makes sure the apps run perfectly with the latest and greatest Docker versions on the chosen platform, hardware, infrastructure, and OS

In the second half of the chapter, we'll be covering the Azure Container Service, which makes it simple to create, configure, and manage clusters of virtual machines that provide the support to run containerised applications. It allows us to deploy and manage containerised applications with Microsoft Azure. It also supports the various Docker orchestration tools, such as DC/OS, Docker Swarm, or Kubernetes as per user choice.

In this chapter, we will cover the following topics:

- Architecture of **Amazon EC2 Container Service** (**Amazon ECS**)
- Troubleshooting AWS ECS deployment
- Updating the Docker containers in the ECS cluster
- Architecture of the Microsoft Azure Container Service
- Troubleshooting the Microsoft Azure Container Service
- Docker Beta for AWS and Azure

Architecture of Amazon ECS

The core architecture of Amazon ECS is the cluster manager, a backend service which handles the task of cluster coordination and state management. On top of the cluster manager sits the scheduler manager. They are decoupled from each other, allowing customers to build their own scheduler. The pool of resources includes CPU, memory, and the networking resources of Amazon EC2 instances partitioned by containers. Amazon ECS coordinates the cluster through the open source Amazon ECS container agent running on each EC2 instance, and does the job of starting, stopping, and monitoring containers as requested by the scheduler. In order to manage a single source of truth: EC2 instances, task running on them and containers and resources utilized. We need the state to be stored somewhere, which is done in the cluster manager key/value store. To be robust and scalable, this key/value store needs to be durable, available, and protect against network partitions and hardware failures. To achieve the concurrency control for this key/value store, a transactional journal based data store is maintained to keep record of changes to every single entry. The Amazon ECS cluster manager has opened a set of APIs to allow users to access all the clustered state information stored in the key/value store. Through the `list` command, customers can retrieve the cluster under management, running tasks, and EC2 instances. The `describe` command can help to retrieve details of specific EC2 instances and the resources available with them. Amazon ECS architecture delivers a highly scalable, available, and low latency container management solution. It is fully managed and provides operational efficiency, allowing customers to build and deploy applications and not think about clusters to manage or scale:

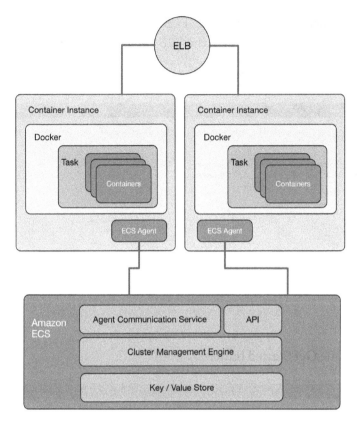

Amazon ECS architecture

Troubleshooting – AWS ECS deployment

An EC2 instance can be deployed manually and Docker can be configured on it, but ECS is a group of EC2 instances managed by ECS. ECS will take care of deploying Docker containers across the various hosts in a cluster and integrating with other AWS infrastructure services.

In this section, we'll be covering some of the basic steps to set up ECS on AWS, which will help to troubleshoot and bypass basic configuration errors:

- Creating an ECS cluster
- Creating an ELB load balancer

- Running Docker containers in the ECS cluster
- Updating Docker containers in the ECS cluster

1. Launch the **EC2 Container Service** listed under **Compute** from the AWS Console:

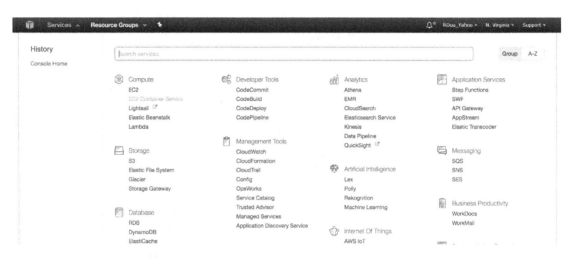

2. Click on the **Get Started** button:

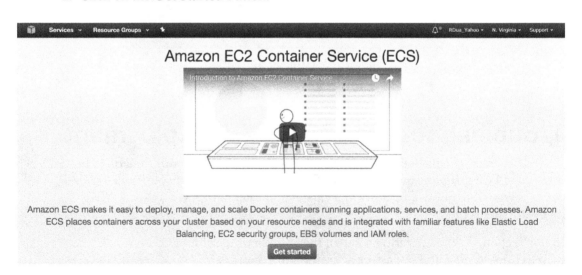

Amazon ECS makes it easy to deploy, manage, and scale Docker containers running applications, services, and batch processes. Amazon ECS places containers across your cluster based on your resource needs and is integrated with familiar features like Elastic Load Balancing, EC2 security groups, EBS volumes and IAM roles.

3. On the next screen, select both options: deploy a sample application, create, and manage a private repository. A private repository is created for the EC2 service and secured by AWS. It requires an AWS login to push images:

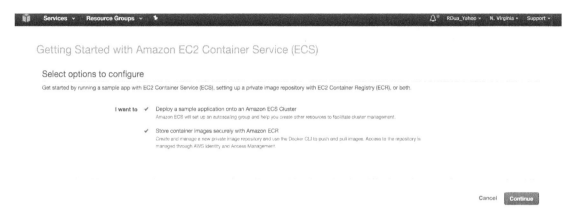

4. Provide the repository name, and we'll be able to see the repository address where container images need to be pushed being generated:

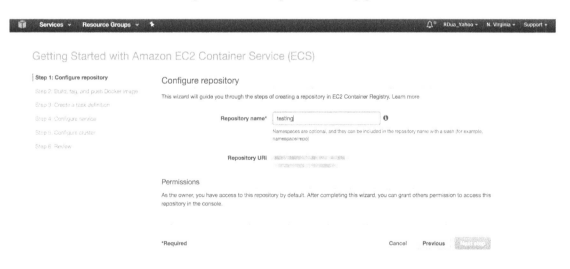

5. The next screen shows some of the basic Docker and AWS CLI commands to push the container images to the private repository, as the following shows:

Install AWS CLI with the help of the `pip` package manager:

```
$ pip install awscl
```

Use the `aws configure` command and provide an AWS access key ID and AWS secret access key to log in:

```
$ aws configure
AWS Access Key ID [None]:
AWS Secret Access Key [None]:
Default region name [None]:
Default output format [None]:
```

Get the `docker login` command to authenticate the local Docker client to the private AWS registry:

```
$ aws ecr get-login --region us-east-1
docker login -u AWS -p
Key...
```

Use the link which is generated as the output of the preceding command which will configure the Docker client to work with the private repository deployed in AWS:

```
$ docker login -u AWS -p Key...
Flag --email has been deprecated, will be removed in 1.13.
Login Succeeded
```

Now we'll tag the nginx basic container image with the AWS private repository name in order to get it pushed to the private repository:

```
$ docker images
REPOSITORY    TAG      IMAGE ID      CREATED     SIZE
nginx         latest   19146d5729dc  6 days ago  181.6 MB

$ docker tag nginx:latest private-repo.amazonaws.com/sample:latest

$ docker push private-repo.amazonaws.com/sample:latest
The push refers to a repository [private-repo.amazonaws.com/sample]
e03d01018b72: Pushed
ee3b1534c826: Pushing [==>] 2.674 MB/58.56 MB
b6ca02dfe5e6: Pushing [>] 1.064 MB/123.1 MB
... Image successfully pushed
```

6. After pushing the image to the private Docker repository, we'll be creating a task definition defining the following:

- The Docker images to run
- The resources (CPU, memory, and other) to be allocated
- The volumes to be mounted
- The Docker containers to be linked together
- The command container that should run when it is started
- The environment variables to be set for the container
- The IAM roles the task should use for permission
- Privileged Docker container or not
- The labels to be given to the Docker container
- The port mapping and network, and Docker networking mode to be used for the containers:

7. Advanced container configuration gives us the option to declare the **CPU units**, **Entry point**, privileged container or not, and so on:

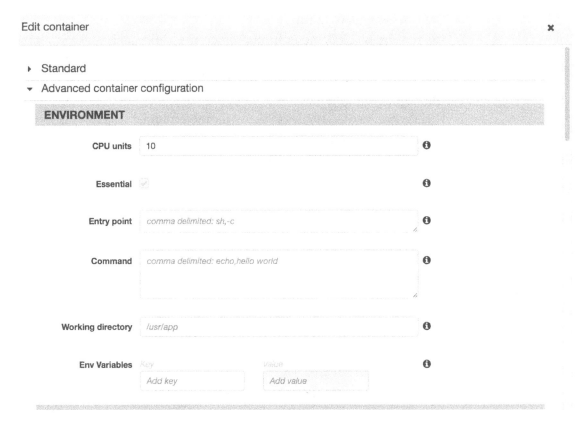

8. In the next step, we'll be declaring the service useful for a task that runs continuously, such as a web service.

This allows us to run and maintain a specified number (desired count) of task definitions simultaneously in the ECS cluster. If any of the tasks fails, the Amazon ECS service scheduler launches another instance and maintains the desired number of tasks in the service.

We can optionally run the desired count of tasks in our service behind a load balancer. Amazon ECS allows us to configure elastic load balancing to distribute traffic across the tasks defined in the service. The load balancer can be configured as an application load balancer, which can route requests to one or more ports and makes decisions at the application layer (HTTP/HTTPS). A classic load balancer makes decisions at the transport layer (TCP/SSL) or application layer (HTTP/HTTPS). It requires a fixed relationship between the load balancer port and container instance port:

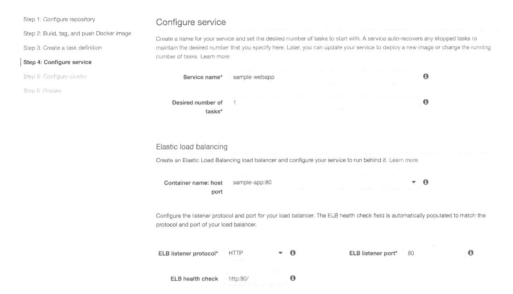

9. In the next step, configure the cluster, which is a logical grouping of EC2 instances. By default, we'll be defining `t2.micro` as an EC2 instance type and the current number of instances as `1`:

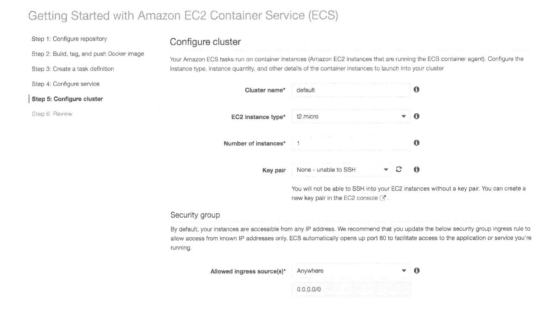

10. Review the configuration and deploy the ECS cluster. After the cluster is created, click on the **View Service** button to see details about the service:

11. Click on the EC2 container load balancer to get the publicly accessible service URL:

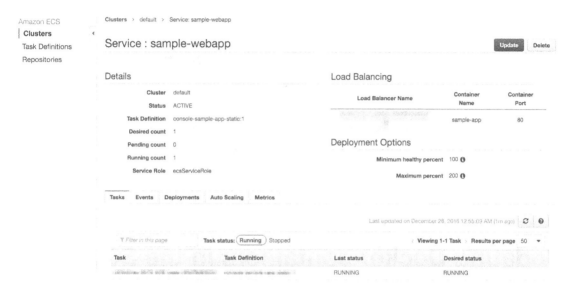

12. In the description of the load balancer, DNS name is the URL to access the service from the Internet:

13. The Welcome to nginx page can be seen as we access the load balancer public URL:

Welcome to nginx!

If you see this page, the nginx web server is successfully installed and working. Further configuration is required.

For online documentation and support please refer to nginx.org. Commercial support is available at nginx.com.

Thank you for using nginx.

Updating Docker containers in the ECS cluster

We have the Docker container running in the ECS cluster, so now, let's walk through a scenario where both the container and the service need to be updated. Usually, this happens in a continuous delivery model, where we have two production environments; the blue environment is the older version of the service and is currently live, to handle users' requests. The new release environment is termed the green environment, which is in the final stage and will be handling future incoming requests from users as we switch from the older version to the newer one.

The blue-green deployment helps to give a rapid rollback. We can switch the router to the blue environment if we face any issues in the latest green environment. Now, as the green environment is live and handling all the requests, the blue environment can be used as a staging environment for the final testing step of the next deployment. This scenario can easily be achieved with the help of **Task definitions** in ECS:

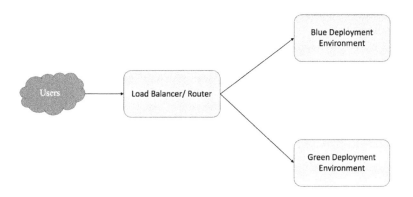

Blue-green deployment environment

1. The new revision can be created by selecting the ECS task created and clicking on the **Create new Task Definition** button:

2. In the new definition of the task, we can attach a new container or click on the container definition and update it. *Advanced container configuration* can also be used to set up the *Environment Variables*:

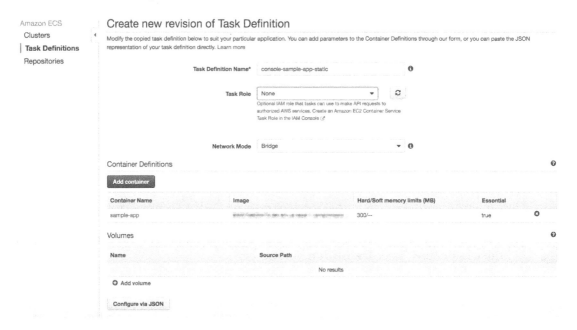

3. After creating the latest task, click on **Actions** and then click on **Update Service**:

4. The **console-sample-app-static:2** will update the **console-sample-app-static:1** and various options, including number of tasks and auto scaling options, are provided on the next screen:

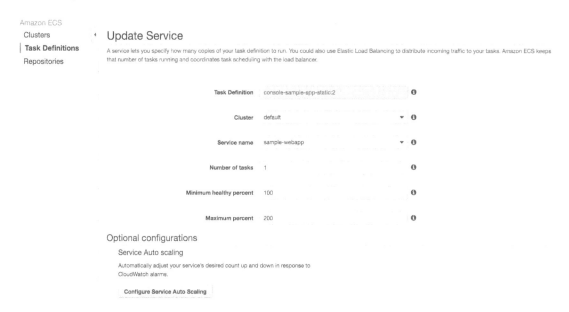

The auto scaling group will launch, including the AMI, instance type, security group, and all other details used to launch the ECS instance. Using the scaling policy, we can scale the cluster instances and services, and safely scale them down as demands subside. The availability zone aware ECS scheduler manages, distributes, and scales the cluster, thus making the architecture highly available.

Microsoft Azure container service architecture

Azure is one of the fastest growing infrastructure services in the market today. It supports scale-on-demand and the ability to create hybrid environments, and big data with the help of Azure Cloud Services. The Azure Container Service provides deployment of open source container clustering and orchestrating solutions. With the help of the Azure Container Service, we can deploy DC/OS (Marathon), Kubernetes, and Swarm based container clusters. The Azure portal provides a simple UI and CLI support to achieve this deployment.

Microsoft Azure is officially the first public cloud to support mainstream container orchestration engines. Even the Azure Container Service engine is open sourced on GitHub (`https://github.com/Azure/acs-engine`).

This step enables developers to understand the architecture and run multiple orchestration engines directly on the vSphere Hypervisor, KVM, or HyperV. The **Azure Resource Manager** (**ARM**) templates provide the basis of the cluster deployed via the ACS APIs. The ACS engine is built in Go, which enables users to combine different pieces of configuration and build a final template that can be used for deploying a cluster.

The Azure container engine has the following features:

- Orchestrator of your choice, such as DC/OS, Kubernetes, or Swarm
- Multiple agent pools (availability set and virtual machine set)
- Docker cluster size up to 1,200:
- Supporting custom vNET

The Azure Container Service is primarily built with DC/OS as one of the critical components, and implementation is optimized for easy creation and usage on Microsoft Azure. ACS architecture has three basic components: Azure Compute to manage the VM health, Mesos for container health management, and Swarm for Docker API management:

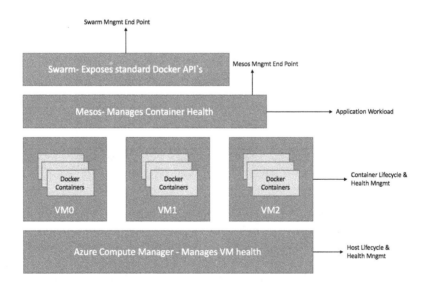

Microsoft Azure container architecture

Troubleshooting – The Microsoft Azure Container Service

In this section, we'll be looking at how to deploy a Docker Swarm cluster and provide orchestrator configuration details in Microsoft Azure:

1. We need to create an RSA key, which will be requested in the deployment steps. The key will be required to log in to the deployed machines post installation:

   ```
   $ ssh-keygen
   ```

   Once generated, the keys can be found in `~/root/id_rsa`

2. Click on the **New** button in your Azure account portal:

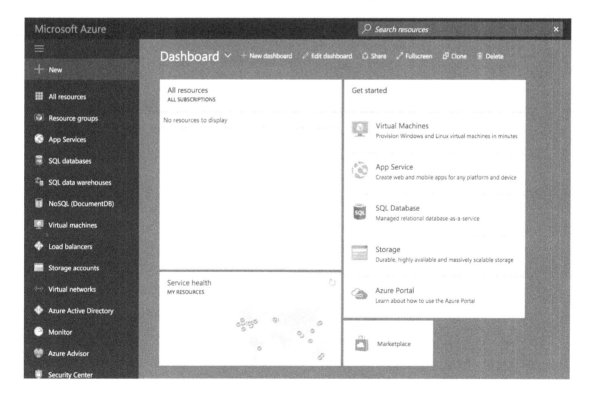

3. Search for the **Azure Container Service** and select it:

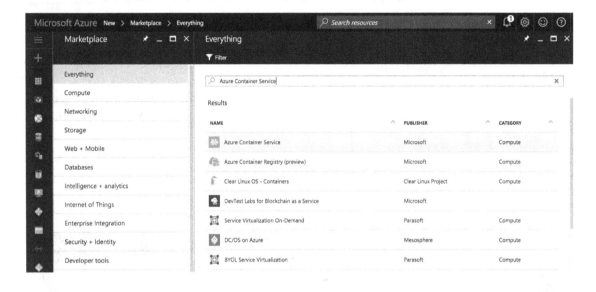

4. After this step, select **Resource Manager** as the deployment model and click on the **Create** button:

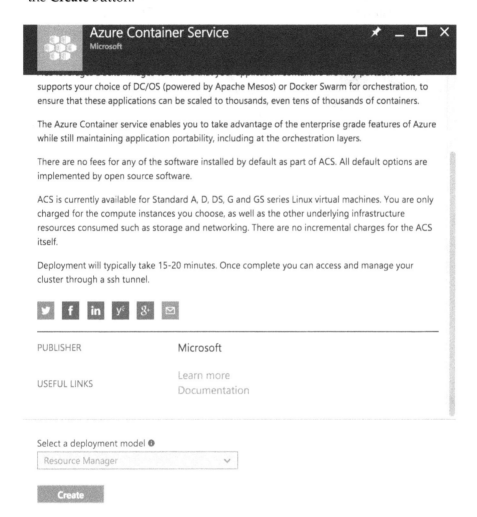

5. Configure the basics settings page, the following details are required: **User name**, which will be administrator for the virtual machines deployed in the Docker Swarm cluster; the second field is to provide the **SSH public key** we created in the step 1; and create a new resource group by specifying the name in the **Resource Group** field:

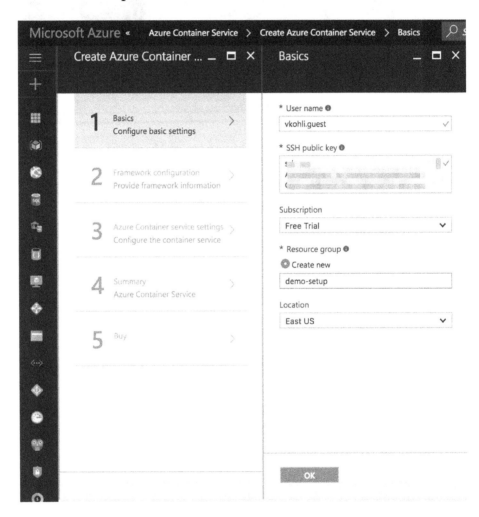

6. Select the **Orchestrator configuration** as **Swarm**, **DC/OS**, or **Kubernetes**, as required:

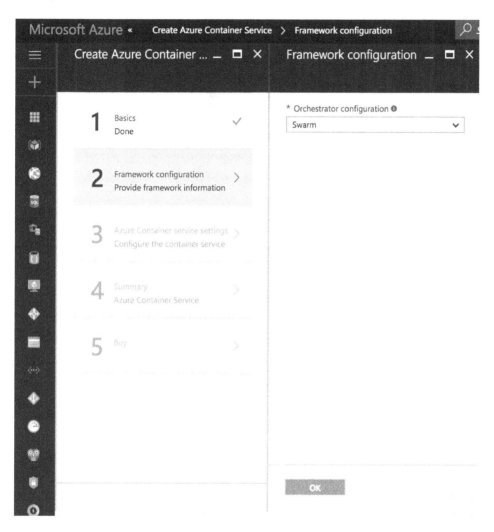

7. In the next step, provide the orchestrator configuration, **Agent count**, and **Master count** for this deployment. Also, the DNS prefix can be provided as `dockerswarm` or as required:

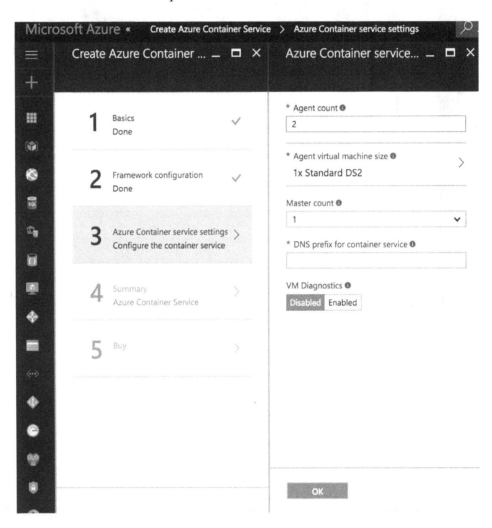

8. Check the **Summary**, and once validation is passed click on **OK**. On the next screen, click on the **Purchase** button to go ahead with the deployment:

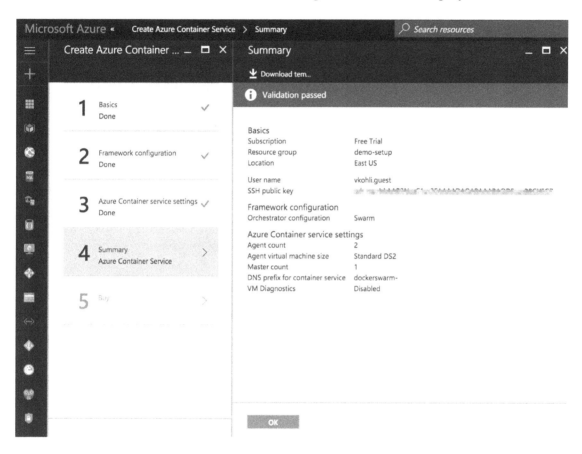

9. Once the deployment has started, the status can be seen on the Azure primary **Dashboard**:

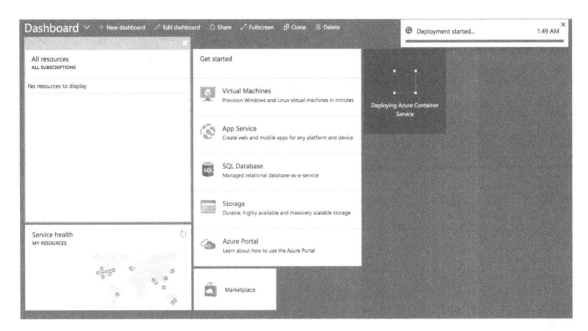

10. Once the Docker Swarm cluster is created, click on the swarm-master from the Docker Swarm resources shown on the dashboard:

11. In the **Essentials** section of the swarm-master, you'll be able to find the DNS entry, as shown in the following screenshot:

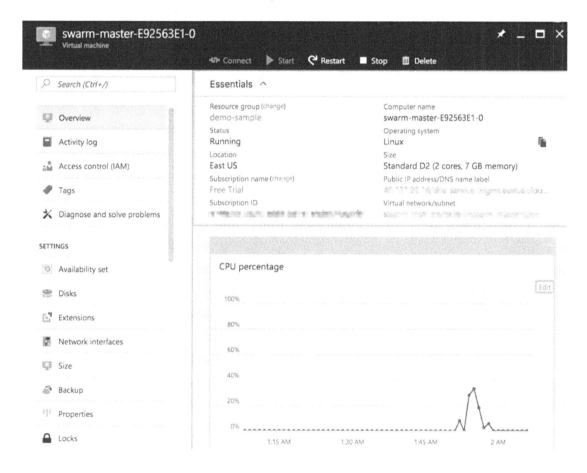

The following is the command to connect via SSH to the swarm-master:

```
ssh <DNS_FROM_FIELD> -A -p 2200 -i <PUB_FILE_LOCATION>
```

Once connected to the master, basic Docker Swarm commands can be executed, and container deployment can be done on the Swarm cluster deployed on Microsoft Azure.

Docker Beta for AWS and Azure

With the recent release of this service, Docker has made it simple to deploy the Docker engine on AWS and Azure through tight integration with both cloud platforms' infrastructure services. This allows developers to bundle their code and deploy it in production machines, regardless of the environment. Currently, this service is in Beta version, but we have covered a basic tutorial of Docker deployment for AWS. This service also allows you to upgrade Docker versions comfortably within these environments. Even the Swarm modes are enabled in these services, which provides a self-healing and self-organizing Swarm mode for the individual Docker engines. They are also distributed across availability zones.

Docker Beta for AWS and Azure provides the following improvements compared to the preceding approaches:

- Using SSH keys for an IaaS account, for access control
- Easy provisioning of infrastructure load balancing, and dynamic updating, as apps are provisioned in the system
- Secured Docker setups can be done with the help of security groups and virtual networks

Docker for AWS uses the *CloudFormation* template and creates the following objects:

- EC2 instances with auto scaling enabled
- IAM profiles
- DynamoDB Tables
- VPC, subnets, and security groups
- ELB

SSH keys of the AWS region are required to deploy and access the deployed instances. The installation can also be done with the CloudFormation template using the AWS CLI, but in this tutorial, we'll be covering the AWS console based approach:

1. Log in to the console, select CloudFormation, and click on **Create Stack**.
2. **Specify the Amazon S3 template URL**
   as `https://docker-for-aws.s3.amazonaws.com/aws/beta/aws-v1.13.0-rc4-beta14.json`, as follows:

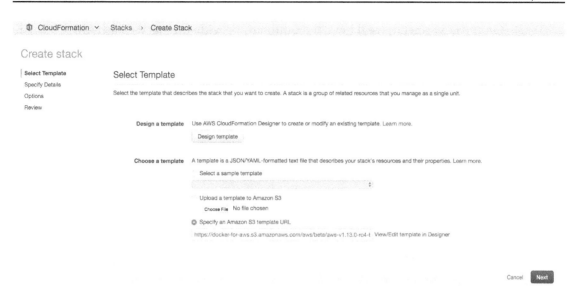

3. On the next screen, specify the stack details, stating the number of Swarm managers and nodes needing to be deployed. The AWS generated SSH key to be used can also be specified:

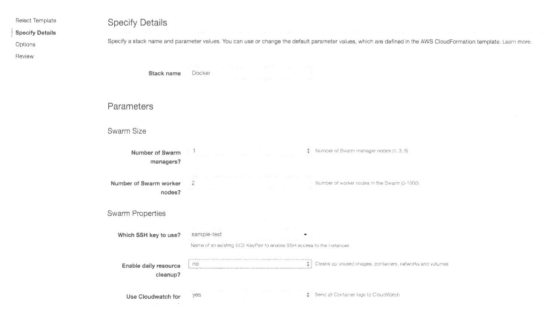

4. On the next screen, we'll have the option to provide tags as well as IAM permission roles:

5. Review the details and launch the stack:

6. The stack will get listed with the status **CREATE_IN_PROGRESS**. Wait till the stack gets fully deployed:

7. Post deployment, the stack will have the status **CREATE_COMPLETE**. Click on it and the deployed environment details will be listed:

The AWS generated SSH keys can be used to SSH into the manager node and administer the deployed Docker Swarm instance:

```
$ ssh -i <path-to-ssh-key> docker@<ssh-host>
Welcome to Docker!
```

The `docker info` command will provide information about the Swarm cluster. The Swarm nodes can be listed using the following command:

```
$ docker info
Containers: 5
 Running: 4
 Paused: 0
```

```
 Stopped: 1
Images: 5
Server Version: 1.13.0-rc4
Storage Driver: overlay2
 Backing Filesystem: extfs
```

```
$ docker node ls
ID                               HOSTNAME                      STATUS
AVAILABILITY  MANAGER STATUS
koewopxooyp5ftf6tn5wypjtd        ip-172-31-37-122.ec2.internal Ready   Active
qs9swn3uv67v4vhahxrp4q24g        ip-172-31-2-43.ec2.internal   Ready   Active
ubkzv527r1r08fjjgvweu0k6t *      ip-172-31-1-137.ec2.internal  Ready   Active
Leader
```

The SSH connection can be made directly to the leader node as well, and a basic Docker container can be deployed:

```
$ ssh docker@ip-172-31-37-122.ec2.internal
```

```
$ docker run hello-world
Unable to find image 'hello-world:latest' locally
latest: Pulling from library/hello-world
c04b14da8d14: Pull complete
Digest:
sha256:0256e8a36e2070f7bf2d0b0763dbabdd67798512411de4cdcf9431a1feb60fd9
Status: Downloaded newer image for hello-world:latest

Hello from Docker!
```

The service can be created for the preceding deployed container as follows:

```
$ docker service create --replicas 1 --name helloworld alpine ping
docker.com
xo7byk0wyx5gim9y7etn3o6kz
```

```
$ docker service ls
ID              NAME          MODE         REPLICAS    IMAGE
xo7byk0wyx5g    helloworld    replicated   1/1         alpine:latest
```

```
$ docker service inspect --pretty helloworld
ID:           xo7byk0wyx5gim9y7etn3o6kz
Name:         helloworld
Service Mode: Replicated
```

The service can be scaled in the Swarm cluster and removed as follows:

```
$ docker service scale helloworld=5
helloworld scaled to 5

$ docker service ps helloworld
ID                 NAME            IMAGE           NODE
DESIRED STATE      CURRENT STATE                   ERROR   PORTS
9qu8q4equobn       helloworld.1    alpine:latest   ip-172-31-37-122.ec2.internal
Running            Running about a minute ago
tus2snjwqmxm       helloworld.2    alpine:latest   ip-172-31-37-122.ec2.internal
Running            Running 6 seconds ago
cxnilnwa09tl       helloworld.3    alpine:latest   ip-172-31-2-43.ec2.internal
Running            Running 6 seconds ago
cegnn648i6b2       helloworld.4    alpine:latest   ip-172-31-1-137.ec2.internal
Running            Running 6 seconds ago
sisoxrpxxbx5       helloworld.5    alpine:latest   ip-172-31-1-137.ec2.internal
Running            Running 6 seconds ago

$ docker service rm helloworld
helloworld
```

Summary

In this chapter, we have covered Docker deployment on public clouds Microsoft Azure and AWS. Both cloud providers provide a competitive container service for customers. This chapter helps to explain the detailed architecture of the AWS EC2 and Microsoft Azure Container Service architecture. It has also covered installation and troubleshooting for all the deployment steps of the container cluster. This chapter has covered the blue-green deployment scenario and how it can be supported in AWS EC2, which is mostly necessary in the case of modern SaaS applications. Finally, we have covered Docker Beta, for AWS and Azure, which was launched recently and provides easy migration of containers from a development environment to a production environment as they are same. Container-based applications can be easily deployed and scaled with Docker Beta, as this service is very well coupled with the IaaS of the cloud providers.

Index

Node.js 10

O

offline key 115
OpenFlow 183
OpenvSwitch (OVS)
  about 183
  configuring, to work with Docker 183
OS containers
  about 8, 9
  versus application containers 10, 11
overlay network
  about 173
  configuring, with Docker Engine swarm node 177
OVS multiple host setups
  troubleshooting 186
OVS single host setup
  troubleshooting 184

P

persistent volume (PV) 210
persistent volume chain (PVC) 210
PHPUnit 79
pod 192
private Docker registry
  about 109
  advantages 109
privileged containers
  about 134
  troubleshooting tips 135
production environment
  Kubernetes, deploying in 207, 208, 209
Project Calico 175
Puppet
  about 138
  use cases 138

R

registry, of Docker images
  reference 51
replication controller 192
role-based authentication (RBAC) 121

S

sandbox 167
scratch repository
  used, for building images 57
SELinux 134
service-in-a-box container 19
Service-Oriented Architecture (SOA) 84
sharding 94
shared volume 16
signed images 115
SSH access
  restricting, from one container to another 161
standardization 50
super-privileged container 136
supermin 57
sysdig
  about 76
  advanced installation 77
  reference 76
  single step installation 77
  used, for debugging images 76

T

tagging key 115
test environment containers 18
The Updated Framework (TUF) 115
three-tier web application
  building 88
tools, to help with Docker networking
  Flannel 174
  Project Calico 175
  Weave 175

U

UFS
  basics 227
  issues 228
  terminology 228
Unikernels
  about 20, 21
  adoption, in Docker toolchain 22
  benefits 20
Universal Control Plane (UCP) 120
Universal TUN/TAP device 174

www.ingramcontent.com/pod-product-compliance
Lightning Source LLC
Chambersburg PA
CBHW060521060326
40690CB00017B/3347

* 9 7 8 1 7 8 3 5 5 2 3 4 4 *